Interdisciplinary Environmental Studies

Interdisciplinary Environmental Studies:

A Primer

By

Gunilla Öberg

⊛ **WILEY-BLACKWELL**

A John Wiley & Sons, Inc., Publication

This edition first published 2011, © 2011 by Gunilla Öberg

Blackwell Publishing was acquired by John Wiley & Sons in February 2007. Blackwell's publishing program has been merged with Wiley's global Scientific, Technical and Medical business to form Wiley-Blackwell.

Registered office: John Wiley & Sons Ltd, The Atrium, Southern Gate, Chichester, West Sussex, PO19 8SQ, UK

Editorial offices: 9600 Garsington Road, Oxford, OX4 2DQ, UK
      The Atrium, Southern Gate, Chichester, West Sussex, PO19 8SQ, UK
      111 River Street, Hoboken, NJ 07030-5774, USA

For details of our global editorial offices, for customer services and for information about how to apply for permission to reuse the copyright material in this book please see our website at www.wiley.com/wiley-blackwell

Library of Congress Cataloguing-in-Publication Data

Öberg, Gunilla.
   Interdisciplinary Environmental Studies : a primer / Gunilla Öberg.
      p. cm.
   Includes index.
   Summary: "Interdisciplinary work is to a large extent a question of entering the unknown, an adventure with exciting and endless opportunities" – Provided by publisher.
   ISBN 978-1-4443-3686-3 (hardback) – ISBN 978-1-4443-3687-0 (pbk.)
1. Interdisciplinary research.   I. Title.

   Q180.55.I48O24 2010
   304.2′8072–dc22

                          2010023312

A catalogue record for this book is available from the British Library.

This book is published in the following electronic formats: eBook 9781444328493; Wiley Online Library 9781444328486

Set in 10/13 pt Rotis Semisans by Toppan Best-set Premedia Limited
Printed and Bound in Singapore by Markono Print Media Pte Ltd

1   2011

Dedication

To Margareta and Bengt – for teaching me to question

Contents

Foreword

The publication of a primer for interdisciplinary environmental studies comes at a time of three important developments: the emergence of a growing number of resources for students and others embarking on interdisciplinary research, the escalation of "transacademic" or trans-sector "transdisciplinary" approaches, and increased attention to questions of quality. Bridging the divides of disciplinary worldviews, qualitative and quantitative approaches, and theory and practice, Gunilla Öberg focuses on the production of written texts in the environmental field, informed by literature on interdisciplinary studies and experience in environmental and sustainability programs in Sweden and Canada.

The book places learning at the heart of the research process in an interrogative approach that fosters critical thinking and cultivation of the "reflective doer." Inclusion of reflections by students and others model the process and may be used both independently and in coursework. The overall structure of sign-posting with numerous examples and exercises guides readers through the maze of framing a study, grounding it in relevant literature, analyzing evidence, gauging interest and relevance, interacting with others in the academy and in society, making creative use of disciplinary differences, establishing common ground, and achieving credibility on both disciplinary and interdisciplinary ground. Öberg also answers frequently asked questions about which methods and research traditions students are most likely to utilize, how to anchor written texts, and how to communicate the aims of a study. Grappling with the challenges of complexity, contextualization, and interpretation can be daunting, let alone the wide range of pragmatic, conceptual, epistemological, and ontological contours of research. Interdisciplinary studies, Öberg notes in her introduction, is a journey into the unknown. Yet, her book demonstrates, it need not be a mystery.

Julie Thompson Klein
Wayne State University,
Detroit, MI, USA

Preface

How do we tackle the complex challenges of the future? What knowledge is needed and what kind of skills? This book has been borne out of uncountable dialogues with students, colleagues, friends and family pondering these questions. As professor and supervisor, the question I constantly struggle with is "What should we teach and how?" We clearly need to increase our ability to integrate various types of knowledge. Many are those who have written on these issues and it is certainly not because of lack of deep insights that we still struggle with the challenges of interdisciplinary studies. My sense is that there is a need for hands-on pragmatic advice, a feeling that has been with me and continued to grow ever since I wrote my bachelor thesis on earthworms and environmental monitoring some twenty years ago. So I decided to write a 'primer' to interdisciplinary studies; the word has many meanings of which one is catalyst that initiates complex actions beyond itself. My hope is that this book will help you get started – and encourage you to keep going.

Teachers and colleagues can inspire both through the quality of their teaching and by standing as fully-formed examples of brilliant scholars. When we begin to write, supervisors, colleagues and friends push us to do our best, inspiring us and giving us confidence. Many have guided me along the way and it is impossible to thank all to whom I owe gratitude. From the beginning of my thinking about interdisciplinarity: my undergraduate supervisor, Dr. Jan Landin, for being a warm and dedicated teacher driven by a deep passion for ecology that I would often remember as I began to work with those equally passionate about their own disciplines. My graduate supervisor, Dr. Anders Grimvall, for stressing the necessity of asking the right questions (and of questioning the questions) and pushing my writing skills. Graduate student colleagues and friends in the communal lunch team for discussions on knowledge and the role of science in society, especially Drs. Johan Hedrén and Karin Sundblad. Research partners: Dr. Karin Bäckstrand for opening my eyes to qualitative studies; Drs. Roger Pielke Jr, Merle Jacobs and Eva Lövbrand for lively and thoughtful discussions on science-policy interaction and Drs. Madelaine Abrandt-Dahlgren and Victoria Wibeck for insights into the theory and practice of Problem Based Learning and Focus Group methodology. Colleagues in the Environmental

Science Program, especially Drs. Thomas Achen and Per Sandén for fruitful discussions on teaching and learning across "the Gap". My chlorine research group, most notably Drs. David Bastviken and Teresia Svensson, for helping me remember that interdisciplinary challenges may also be pertinent in "pure" science studies. My passionate and colorful colleagues at UBC: Dr. Terre Satterfield for sharing splendid practical exercises and for thoughtful, inspiring and fun discussions on quality assessment and the nature of evidence; Dr. John Robinson for helping me to see the broader picture; Dr. Douw Steyn for in-depth and careful reading of an early version; Dr. Hadi Dowlat Abadi for insightful, constructive and thought-provoking comments and propositions; Dr. Kai Chan for sincere and carefully thought-through discussions on "usefulness".

When we teach and supervise, we benefit from students who, as junior members of the academy, bring openness and excitement to the problems we are wrestling with in our work. I am deeply indebted to the countless number of students who have provided feedback while reading earlier versions of the manuscript. Special thanks to: the students from the first cohorts in the Environmental Science Program at Linköping University for discussions on quality assessment; the group of graduate students at UBC who took the time to discuss and dissect each chapter of an early version, which led to a total revamping of the entire manuscript: Meg O'Shea, Negar Elmieh, Tom Berkhout, Lara Hoshizaki, Conor Reynolds, Jack Teng, Sonja Klinsky and Nicole Dusyk; and finally, my warm thanks to the four students who have been "test-driving" the questions of Chapters 4 and 5: Natalie Ban, Charlie Wilson, Shannon Hagerman and Jane Lister, who also have acted as a sounding board on the hills or over a glass of wine. Many warm thanks to Elizabeth Maurer for helping me writing English good – she edited the text, skillfully eradicating all "swedicisms" while sensibly allowing my personal style to color the text, and to Andrea Chamberlain for making the cover painting, skillfully capturing the hue and texture I had in mind. I also wish to thank my editor Ian Francis and the team at Blackwell and Wiley; it has been a pleasure to work with you.

Last but not least, my warmest thanks to my loving family: to my parents Margareta and Bengt for encouraging curiosity, creativity and hands-on pragmatism. To my children Anna and Martin for thoughtful questions, insightful responses and for challenging me to stay open-minded. And to Doug, for the book title, for listening and feed-back, and for being more stubborn than I am.

Gunilla Öberg
Vancouver, Canada
August 2010

1 Introduction

Interdisciplinary[1] work is to a large extent a question of entering the unknown, an adventure with exciting and endless opportunities. Since each setting is new and the details cannot be foreseen, students and scholars who work in interdisciplinary environments must be independent and self-driven. This book is therefore more of a guide than a how-to handbook. "High quality" is shaped differently in different disciplines. Rather than letting these differences be a hindrance, you can use them as a springboard for inventive and original work. The governing idea of this book is to facilitate creation of interdisciplinary work by stimulating dialogues on quality and to draw on common-ground-creation processes to find unknown and unexplored territories.

This is first and foremost a book for graduate and undergraduate students about to enter the interdisciplinary world, but my intention is to write a book that also is useful for teachers, supervisors, researchers and editors who are active in interdisciplinary settings.

Drawing on my own experiences and leaning on the literature on inter-disciplinarity, I strive to facilitate the development of "Reflective Doers".[2] The target is mainly students and scholars in the environmental field, who work with issues that involve interaction between and among the human and natural worlds and who consequently may have quite diverse disciplinary backgrounds – spanning from natural sciences, technology and health sciences to the social sciences and not least the humanities.

I strive to use a simple language and as far as possible avoid jargon and specialist terminology. Things that are self-evident for a physicist are alien to an anthropologist and vice versa. For example, a colleague who read an early version of the manuscript asked why I do not use the terms nominal, ordinal, interval and ratio data. The reason I avoid these terms is that even though they are basic knowledge for someone with a background in mathematics or physics, they are impenetrable jargon for everyone else, who consequently would be excluded from the text. So, if you feel that my text occasionally is simplistic, it is likely because I avoid or tread carefully when using terms that to my knowledge are ambiguous in interdisciplinary contexts, common knowledge to you but Kiswahili to someone else.

Now and then I touch upon subjects that are discussed at length in other forums. I try to highlight when I touch upon such fields and in the footnotes I suggest some readings, but I will not engage in these ongoing debates, as it would lead too far from the aim of this book.

In Chapter 2, I delve on "The Gap" between humanities and social sciences on the one hand and natural sciences, technology and medicine on the other. The focus lies on real and perceived gaps that often cause problems and how they may be approached to enable creation of common ground. In Chapter 3, I present a framework that is designed to facilitate the creation of high-quality interdisciplinary work. A key aim of this framework is to empower you to identify, accept, respect and draw upon disciplinary-based cultural differences; in other words, to help you find ways to *use* the differences. In contrast, if you are unaware of the differences among the academic cultures you encounter, it is likely that they will impede your work. The framework illuminates three dimensions of interdisciplinary quality:

1) integration of elements from different disciplines.

2) interaction with organizations and individuals outside academia.

3) rigour from an academic point of view.

In Chapter 4, I discuss the first part of the framework: specific demands on different types of interdisciplinary work with a focus on disciplinary boundaries within academia. Chapter 5 focuses on the second part of the framework: boundaries between society and academia.[3] In Chapters 6 to 11, I dive into the sub-components of the third part of the framework: academic rigour. These components are valid for anyone conducting academic work and this part of the framework thus applies to disciplinary as well as interdisciplinary environments. While disciplinary students may learn how to conduct rigorous studies through apprenticeship rather than through conscious analysis and reflection, interdisciplinary scholars need to address these issues consciously and explicitly – since beauty comes in many forms. In Chapter 12, finally I discuss quality assessment in light of the rest of the book and how hierarchies in academia, as well as arrogance and snobbishness, hinder collaboration.

Challenges and opportunities

To become a student in an interdisciplinary context can be confusing. Students and scholars who are active in a disciplinary environment use that context as a springboard when they plan a study, collect data, analyse, read and write. It is as a rule more challenging to work in an interdisciplinary environment, since such settings often embrace differing and in some respects incommensurable academic cultures.[4]

To draw on the strength of interdisciplinary work you need to manage differences among academic cultures. If you work with extra-academic partners, you also need to manage differences between academia and society. A well-known route to success is to create a climate that stimulates awareness of interdisciplinary opportunities, not least as it will help you to identify your own viewpoints and limitations.[5]

There is a large and growing body of literature on interdisciplinary work.[6] This literature is, unfortunately, not of much help to the newcomer. In general, the focus of this literature is on challenges, barriers and problems, which easily could give you the impression that interdisciplinarity is a "mission impossible". As has been pointed out by Lattuca (2001), the literature on interdisciplinary study offers "a litany of geopolitical metaphors /.../ that creates the impression that academic disciplines are foreign territories and interdisciplinarians, hapless trespassers" (p. 243). Research on interdisciplinary studies is both a large and emerging research field, with its own theories and jargon. The literature is mainly written for and read by those who conduct studies on interdisciplinary work. Paradoxically, one might even say that the field has developed into a discipline firmly situated in the humanities. We, those conducting interdisciplinary work, are (as described in the literature) busy in our own research fields, reading literature, participating in conferences and the like. It is unrealistic to expect that any newcomer has the capacity to be in the forefront of his or her (new or emerging) field of study and at the same time be literate in research on interdisciplinary studies. As pointed out by many before me, an easy-to-read guidebook is desperately needed. However, previous attempts speak more to those studying interdisciplinary work than to those doing it. My audience is those who conduct interdisciplinary work and even though I avoid the insider jargon, I draw heavily on the literature as many of the findings undeniably are extremely helpful for the "doer".

One thing that we learn from the literature is that interdisciplinary work is nothing new and successful interdisciplinary studies have been conducted for as long as disciplinary studies.[7] There have always been scholars and students who challenge and cross traditional boundaries. Highly successful interdisciplinary projects have constantly been carried out and many claim that the most exciting and cutting-edge studies have always been born in such settings. It is true that the literature dwells on the difficulties of interdisciplinarity, but the literature also shows that it certainly is not a mission impossible. On the contrary, interdisciplinary studies hold tremendous opportunities.

On quality

The crux of successful interdisciplinary work is to acknowledge that the basis for judging quality varies among disciplines.[8] Quality is to a large extent achieved by adhering to agreed norms on how things should be done – norms that are handed on by traditions. A number of explicit and implicit norms lie beneath each and every discipline and these norms are, to a considerable extent, mirrored in the text. What is to be told, how to tell it and where the various components should be placed vary from discipline to discipline.

Written texts are the focal point of this book, since they are crucial for most academic work.[9] Fieldwork, oral presentations, seminars, laboratory work, workshops and other types of discussion forums play an indisputable role in academia but it is through the production of texts that we in academia render possible the scrutiny of our studies by our peers. It is through our texts that we disseminate our results to a wider public. It is through texts that findings, discussions and controversies are made available for generations yet to come. Students in most disciplines are therefore trained to write various types of texts.

There is a vast literature that focuses on academic text writing, which is of invaluable help as long as you stay fairly well within the same discipline. The common dogma is often that there is one and only one way to produce credible texts. In contrast, Janet Giltrow elucidates and discusses different academic writing traditions in an eloquent manner. But neither Giltrow, nor anyone else to my knowledge, discusses the opportunities (and challenges) of interdisciplinary contexts. In addition, there are few books that set out to define common criteria of credibility without arrogantly defining it as "research done according to my tradition", thereby dismissing other traditions as "bad" only because the research does not conform with the authors' disciplinary-based interpretation of credible research. With this book, I wish to make it easier to identify common ground as well as differences by facilitating dialogue and collaboration.

It is crucial to understand that among the scholars you meet there will be those who by their own experience have deep insights into the opportunities and challenges of interdisciplinary work without being familiar with the literature on interdisciplinarity, those who have no personal experience of interdisciplinary work but have gained deep insights through the interdisciplinary literature and others who will be unfamiliar with anything outside the traditions of their own discipline (Figure 1). You must learn to navigate such a landscape. You must acquire the ability to recognize credibility outside your own area of competence and you must learn how to help others see that your work is credible. You must learn to handle the fact that others will not intuitively understand on what basis your work is credible.

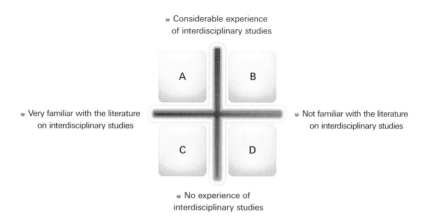

Figure 1 An illustration of the landscape of interdisciplinary competence along the dimensions of own experience and familiarity with literature on interdisciplinarity.

Background

This book draws upon work conducted by the team developing the Environmental Science Program[10] at Linköping University (LiU), Sweden, in combination with ongoing development of a quality assessment framework for graduate studies at the Institute for Resources, Environment and Sustainability (IRES) at the University of British Columbia (UBC), Vancouver, Canada.[11] An early version of the book has been published in Swedish[12] and the third part of the framework has been presented in a paper published in *Higher Education*.

The Environmental Science Program at LiU is an interdisciplinary bachelor and master's programme, which was launched in 1998. The development, chaired by myself, encompassed lengthy discussions among a group of scholars with differing academic backgrounds on the curriculum to identify key skills and abilities necessary for future environmental professionals.

The basic level courses in the programme integrate elements from the humanities, the social sciences, natural sciences and engineering and make use of both qualitative and quantitative approaches. All those involved in the development of the programme, as well as the majority of the tutors and supervisors, had experience with interdisciplinary work (but no one was at that time familiar with the literature, i.e. we were all firmly placed in square B in Figure 1). The development of the basic courses was both time and energy consuming but they were launched with only minor communication problems among the teachers. In contrast, supervision and examination of bachelor and master's theses surfaced as an intricate problem. Starting in the fall of 2001, all supervisors participated in a number of seminars focusing on evaluation criteria. Concrete examples were used each time to initiate our discussions, often one or two theses in which the examiner and the supervisor did not agree on the quality of the work. I acted as facilitator of the discussions, took notes and revised the notes into a working document that was used as a basis for the evaluation procedure. The third part of the framework presented in Chapter 3, which focuses on academic rigour (Q6–Q10), is a revised version of the framework that has been used in the Environmental Science Program at LiU since the spring of 2002.

IRES, at UBC, Vancouver, Canada is an interdisciplinary research institute that is the home of a major graduate programme with over 100 students. IRES strives to foster sustainable futures through integrated research and learning about the linkages among human and natural worlds, to support decision-making on local to global scales. The unit has 12 core academic staff and about 40 associates in 9 of UBC's 12 Faculties and the research is for the most part conducted in close collaboration with non-academic partners. In October 2006, I became the Director of IRES. Quality assessment soon surfaced as an issue of topical interest among students as well as supervisors and committee members, to a large extent driven by the fact that the latter often represent rather diverse academic cultures. In the development of functional tools for planning, conducting and assessing of master's and doctoral theses at IRES, the academic quality assessment questions (Q6-10) were complemented with two additional sets of questions: how to integrate elements from different disciplines (Q1-3) and how to interact with society outside academia (Q4-5).

A note on terminology

Leafing through the literature shows that the term "interdisciplinary" is used to describe quite different activities.[13] The term "interdisciplinary" is, for example, used to signal work that goes beyond disciplines as well as interaction with the broader society outside academia. Unfortunately, the terminology in this field is convoluted and I therefore describe some key concepts and explain how I use them.

Interaction within academia

The fundamental reason people conduct interdisciplinary studies is that a disciplinary approach fails or is insufficient in creating an understanding of the question in focus.[14] I like the way Donald T. Campbell[15] describes interdisciplinary work and how it differs from monodisciplinary and multidisciplinary work. The figure below is inspired by his ideas.

In monodisciplinary work, scholars specialize in areas within the borders of their disciplines. In multidisciplinary work, scholars also specialize in areas within the borders of their disciplines but communicate and interact actively with scholars from other disciplines. This approach brings about a deepened understanding of contributions made by other disciplines and bringing together various studies provides a multifaceted description of the issue in focus. Multidisciplinarity is usually defined as collaboration among disciplines where each participating discipline remains within its traditional framework, whereas interdisciplinarity usually is used to signify some sort of integration.

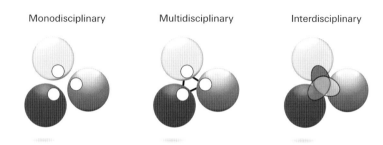

Monodisciplinary   Multidisciplinary   Interdisciplinary

Figure 2 An illustration of the difference between monodisciplinary, multidisciplinary and interdisciplinary work, inspired by the fish-scale model introduced by Donald T. Campbell (2005).

Interaction between academia and the broader society

Sometimes it is argued that interdisciplinary work *must* be conducted in close interaction with stakeholders. This is unfortunate because it creates unnecessary fence-building and confusion. It is clearly possible to conduct advanced interdisciplinary research from a purely academic perspective and there is no reason to exclude such studies when discussing interdisciplinary work. All who strive to use knowledge from more than one discipline face similar challenges. The aim of purely academic research is to create new knowledge. The explicit aim of education and research conducted in close interaction with the broader society outside academia is to produce socially-robust knowledge,[16] which is, at the same time, reliable from a scholarly perspective (Quadrants C and D of Figure 3). The key difference between pure academic research and this latter type of education and research is the extent of interaction with the larger society outside academia.

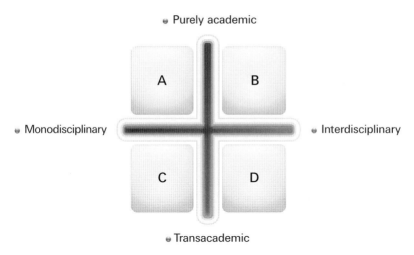

Figure 3 An illustration of the distinction between interdisciplinary and transacademic work. The horizontal axis depicts increasing use and integration of knowledge from more than one academic discipline. The vertical axis depicts the degree of interaction between academia and the broader society.

Interdisciplinary and transacademic work

Research and education, which aim at enhancing our understanding of complex contemporary phenomena, often require the use and integration of knowledge from more than one discipline. Often, it also requires the involvement of extra-academic participants in the research process.

Research in complex areas of topical societal interest often demands the combination of the two. The opportunities (and challenges) in using knowledge from various disciplines are undeniably different from those of close interaction with society outside academia, albeit intertwined.[17] For those conducting interdisciplinary studies, there is reason to separate the two, as such a separation will facilitate the planning, conducting and assessment of such studies.

Education and research conducted in close interaction with the broader society outside academia is sometimes called truly interdisciplinary, issue-driven interdisciplinarity or transdisciplinary. These terms signal that "disciplines" are central to the phenomenon described. This is unfortunate since none of them signals the core focus of this type of activity: the interaction between academia and the larger society. Kinihide Mushakoji introduced the term "transacademic" in 1978. In contrast to the other, more commonly used terms, "transacademic" draws attention to academia, rather than its constituent disciplines, and the term signals activities that both involve and go beyond academic activities. So far, I have been using "interdisciplinary" in its broadest sense, denoting academic activities that go beyond the disciplinary. To facilitate a separation of opportunities (and challenges) related to "using various disciplines" from those related to "interaction with society", I hereafter use "interdisciplinary" to denote the former and "transacademic" to denote the latter. More about interdisciplinary work in Chapter 4 and transacademic work in Chapter 5. The book as a whole focuses on interdisciplinary work, which may or may not be transacademic (quadrants B and D in Figure 3).

Notes

1   I use interdisciplinarity and interdisciplinary work in their broadest senses, denoting academic activities that go beyond the disciplinary. A discussion on terminology follows in the section "A Note on Terminology".
2   More about "reflective doers" in Chapter 4.
3   See, for example, Michael Field *et al.* 1994.
4   See, for example, John Bradbaer 1999; Andrew Barry *et al.* 2008.
5   Peter Bohm 1996; Lisa Lattuca 2001; Angela M. O'Donnell and Sharon J. Donnell 2005; Robert Frodeman *et al.* 2010.
6   For summaries, see, for example, Julie Thompson Klein 1990; Lisa Lattuca 2001; Liora Salter and Alison Hearn 1996; Sharon J. Derry *et al.* 2005; Robert Frodeman and Carl Mitcham 2003; Alan Repko 2008; Robert Frodeman *et al.* 2010.
7   See, for example, Andrew Barry *et al.* 2008 and references therein.
8   For an informed, insightful, easy-to-read and entertaining discussion on different academic styles, see Janet Giltrow, *Academic Writing*, 3rd edn, 2002.

9   My intention is not to comment upon or discuss the production of knowledge from an ontological or epistemological perspective. Instead, I take a practical and pragmatic approach with the aim of enabling education and research efforts across traditional academic borders by elucidating common understandings of credible academic work.

10   http://www.liu.se/tema/miljo/

11   http://www.ires.ubc.ca

12   Gunilla Öberg 2009.

13   For a summary of the discussion on terminology, see Lisa Lattuca 2001. For older literature, see Margaret Barron Luszki 1958; Leo Apostel 1972; Roger B. M Cotterell 1979; Heinz Heckhausen 1972; Erich Jantz 1972; Freremont E. Kast and James E. Rosenzweig 1970; Richard Barth and Rudy Steck 1979.

14   Lisa Lattuca 2001; Veronica Boix Mansilla and Howard Gardner 2003; Christopher Heintz *et al.* 2007; Robert Frodeman *et al.* 2010.

15   Donald T. Campbell 2005.

16   Michael Gibbons *et al.* 1994; Helga Nowotny *et al.* 2001.

17   Sherry R. Arnstein 1969; Arnim Wiek 2007; Roger Pielke Jr 2007; Walter *et al.* 2007.

2 Beyond CP Snow

Many of the challenges we face in the 21st century deal with issues that demand a deepened understanding of the interplay between the biophysical world, humans, society and technology. Research and education stand upon a founding structure with rather distinct borders between the different traditions. As a result, the methods, theories, models, concepts and thought styles that are at our disposal today have mainly been developed to explain, understand and predict phenomena and processes *within* the different academic spheres (roughly divided into the humanities, social sciences, natural sciences and technology). These spheres can only to a limited extent help us understand the interaction *between* nature, humans, society and technical systems. A deepened understanding of the interaction between these spheres demands access to theories, methods, models, concepts and thought styles that not only enlighten these spheres but also the interaction among them. We need informed interdisciplinary studies by skilled interdisciplinary scholars.

When speaking about differences among traditions in academia, the concept of "the two cultures" easily comes to mind.[1] Even though CP Snow, when formulating the concept in 1959, referred to the gap between arts and science, the concept has later come to denote The Gap between the humanities and the social sciences on the one hand and the natural sciences and technology on the other.[2] The concept of the two cultures has undoubtedly initiated various interesting discussions, but it has also led to unnecessary and time-consuming fence-building between the parties. This fence-building exacerbates differences and is of no help for those who try to bridge the gap.

The following chapter deals with the common pitching of qualitative and context dependent humanities and social sciences against quantitative and exact natural sciences, as if this was a full description of the academic

Interdisciplinary Environmental Studies: A Primer, 1st edition. © Gunilla Öberg.
Published 2011 by Blackwell Publishing Ltd.

landscape instead of a meagre oversimplification thus hindering exchange of knowledge across various gaps. The following is an attempt to help you move to a more nuanced understanding, to a more fruitful position, making it possible to understand and draw on differences.

Quantitative and qualitative studies

Most people seem aware that quantitative research means counting, measuring and using statistics and the like, whereas qualitative research is less well known beyond the realm of those who practise it. Many appear to believe that qualitative studies are conducted only in the humanities and the social sciences, and in some quarters it is believed that quantitative studies are conducted only within the natural sciences. Both these apprehensions are misunderstandings: there are without doubt qualitative natural science studies as well as quantitative humanities and social science studies; as developed below, both of these are common and widespread. The words "quantitative" and "qualitative" are used differently and given different connotations in different disciplines, which complicate these discussions since many seem to feel that their interpretation is exclusive while others are ill informed. The differences should not be ignored but there are actually quite a few commonalities that cut across most traditions, commonalities which, however, are seldom explored, discussed or used.

Broadly speaking, qualitative and quantitative research are two different methodological approaches. Qualitative studies focus on the inner quality of things, aim at answering questions such as "how" and "why", and deal either with issues that cannot be quantified or issues where quantification has no meaning. Qualitative studies are common in disciplines such as philosophy, sociology, analytical organic chemistry, political science, taxonomy, ecology and history. Quantitative studies focus on measuring things and aim at answering questions such as "how much", "how fast", and "how long". Quantitative studies are common in economics, hydrology, chemistry, physics, sociology, biogeochemistry, geography and political science.

To illustrate what a quantitative study might be, as compared to a qualitative study, I use an example that cuts across the traditional nature-culture gap.

Example 1
Quantitative and qualitative questions

Quantitative question: How many people can this field feed?
To answer this question you could harvest the field, weigh the rice and calculate how many kilogrammes there were per hectare. The number would be more or less exact but would most likely vary between plots and between years. Then you could proceed and analyse the nutritional value of the rice, and estimate how many people the field might feed by combining the harvest information with information on nutrient needs. Another approach would be to estimate how much the rice would be worth on the local, regional or global market and estimate how many persons the income could provide for.

Qualitative question: What signifies good rice?
To answer this question you could interview people about their perception of good rice. The perception of good rice might be that it is has a high nutritional value, is easy to store, tastes good, is glutinous, non-sticky, etc. You could also set out to investigate whether or not a specific chemical compound can be tied to the taste, odour or colour of popular rice or if its popularity might be tied to a combined effect of texture, taste and odour.

Improved understanding and quality

Few researchers are trained in both quantitative and qualitative methods, and even though one might discern a change with an increasing number of younger scholars who are trained in both, most scholars still act in groups that conduct either qualitative or quantitative research. This either-or division has the unfortunate outcome that qualitative studies with semi-quantitative elements far too often contain quantifications of questionable quality and that quantitative studies with semi-qualitative or qualitative elements rarely are credible in the qualitative parts. This situation unfortunately reinforces the division in either-or. Increasing your methodological awareness will help you combine qualitative and quantitative studies in a credible manner.

When qualitative studies contain sloppy quantifications, it appears to be a question of ignorance of basic statistics. I dare say this is true, whether it is studies in the humanities (e.g. based on in depth interviews) or studies in the natural sciences (e.g. based on analytical chemistry). When conducting studies in which you intend to use quantifications or even semi-quantitative estimates (more than or less than, rather than exact numbers), it is advisable to engage someone who has quantitative training, not only to crunch the numbers when the study has been conducted but first and foremost to assist in the design, since a *bad design cannot be fixed with advanced statistics.*

When quantitative studies rooted in the natural sciences contain qualitative sub-studies of human or societal issues, it appears to be a question of what might be called basic methodological ignorance: in spite of the vast amount of well-written literature in the area, a surprising number of researchers carry out interviews, surveys and document studies without even having opened a book on the procedures in question. In contrast, it never ever happens that a person unskilled in a natural science discipline would carry out a study of processes or phenomena in Nature. For example, I do not believe that it is possible to find a historian who has carried out chemical analyses of old ice cores based on the argument that the ice cores are historical records and that historians thus have the proper training to conduct such studies. In contrast, numerous historical studies have been carried out by natural scientists, engineers or medical researchers, solely based on the qualification of being personally involved in the historical events. Personal experience can be a strong asset in research, but it *cannot* replace methodological rigour. Work conducted without proper basic methodological competence generally produces questionable results when examined according to basic criteria for good scholarly work.

It should go without saying that serious interdisciplinary studies demand that all studies be carried out with equal rigour and be based on sound methods. If you plan to conduct a study that entails quantitative elements, you need to be knowledgeable about quantitative design, discuss the design of your study with an experienced quantitative scholar and not least make sure you involve someone with know-how on statistical analyses. In exactly the same way, if you plan to conduct a qualitative study and, for example, use interviews or surveys, you need to learn how to design a study that draws on those methods and to make sure that someone with know-how conducts or supervises the analyses.

Drawing on commonalities

There is a rich literature on qualitative methods within the humanities and social sciences, but this literature explicitly or implicitly excludes qualitative natural sciences.[3] The types of studies, which are called qualitative within the natural sciences, are thus not embraced in this literature. There are undeniably large differences between qualitative studies in the social sciences and humanities and qualitative studies in the natural sciences.[4] These differences should not be neglected. But there are also a number of commonalities that it should be possible to draw upon. This is seldom done.

Qualitative natural sciences focus on the inner quality of something, and a great number of the general questions that are dealt with in the qualitative method literature are just as relevant for natural science studies as for the social sciences and humanities. We all carry images of how the world is composed, and these images do, of course, influence our way of conducting research.[5] Textbooks on qualitative methods commonly claim, almost in passing, that this matters only for the social sciences and the humanities and that the knowledge in those books therefore is irrelevant to other fields. I disagree as there clearly is a lot to gain if one manages to elucidate and draw upon commonalities among different qualitative disciplines – across the nature-culture gap. In order to make the knowledge from the general qualitative methods literature applicable in a wider context, these texts need to include rather than exclude the natural sciences, medicine and engineering. While waiting for the literature on qualitative methods to become more inclusive, you can re-write them in your mind by ignoring the limits set by the author and identifying examples from the excluded traditions. Below, I illustrate how knowledge from the qualitative methods literature can be helpful for scholars in field-based observational studies and laboratory-based experimental studies.

There is a relatively clear demarcation between the field-based and the laboratory-based natural science traditions, where the former mainly depend on observation and the latter depend on experimentation (Chapter 8). Qualitative as well as quantitative studies are conducted in both areas. Most of the field-based natural science traditions originate from natural history, where Sweden can pride itself in having produced scholars such as Elias Fries and Carolus Linneaus. Natural history stems from the idea that observation makes it possible to increase our understanding of the biophysical world. In the 18th century, Nature was depicted through detailed drawings in combination with illustrative descriptions. The focus was on describing the inner character of various phenomena with the aim of furthering the understanding of relationships, to explain rather than predict.[6] During the late 17th and early 18th centuries, natural history was closely related to natural theology; the idea was that through the understanding of Nature one would be able to form a deeper understanding of God, of one's own existence and its inner meaning. Descriptions of Nature were mainly qualitative but contained some quantitative elements: annotations of temperature, precipitation, numbers of bird or plant species, etc. Various field-based natural sciences have developed from this tradition, such as hydrology, ecology, meteorology, natural geography, climatology and geology, in addition to systematics (how various species are related). Most of these traditions still entail both qualitative and quantitative elements. I use biodiversity as an example where the term "diversity" suggests a quantitative approach. The most common way to describe biodiversity is to measure species abundance. Biodiversity is, however, a complex and intricate concept which, among other things, entails four levels: genes, species, populations and ecosystems. In addition, the term signals both the present situation as well as the ability to withstand or absorb future changes. Biodiversity is thus a concept that is meant to describe an area's inner features, its qualities from a certain perspective. In order to cover the meaning of the concept, various tools have been developed, some of which are based on statistical models. Quantitative tools are thus used to describe a qualitative concept.

The laboratory-based experimental natural sciences rest upon a different heritage as compared with the field-based natural sciences. Among other things, the roots can be traced to alchemy rather than natural history. The aim of alchemy was to figure out how, through manipulation and experimenting, one could achieve a certain goal, as for example creating gold from lead or finding the formula for the universal cure for all diseases. Just as in natural history, there was no clear border between religion and science, at least not until the 19th century.

Among the laboratory-based sciences are the so-called "exact sciences". As can be deduced from the name, the "exact sciences" conduct research in which exact answers are sought, and often possible to find. For example, a specific chemical compound has a specific structure irrespective of the quantity. A compound is either lead or gold. If the structure is changed, the compound is no longer the same. Another example is illustrated in Figure 4, which depicts carbon monoxide and carbon dioxide. Both are gases, but carbon monoxide is comprised of one carbon atom and one oxygen atom, whereas carbon dioxide has two oxygen atoms.

Carbon dioxide is a non-poisonous gas that is essential for photosynthesis and exhaled by humans and many other organisms. Carbon monoxide is also a gas but in many respects differs from carbon dioxide, for example by being poisonous for humans, causing suffocation if inhaled. Identifying a molecule as carbon dioxide is exact, as it either is or is not carbon dioxide.

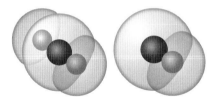

Figure 4 A visual representation of carbon dioxide (CO_2) and carbon monoxide (CO).

At the same time it is a qualitative task, as it is a way to define and identify the intrinsic properties of the gas – its qualities.

Both the field- and laboratory-based natural sciences often use technical equipment that delivers numerical data. Diagrams, numbers, illustrations or equations based on such numerical data are often used to describe various inner qualities. The use of technology, or artefacts as it is often called in the literature, and its role in the production of knowledge is in itself an extensive discussion.[7] Here, I confine the discussion to pointing out that the use of numerical data as *one of several* tools to describe qualitative features most likely is the root of the perception that the qualitative natural sciences are quantitative.

Below, I provide a highly simplified example of how numerical data can be used to describe a qualitative feature. We can distinguish the colour red from blue by the use of visual experience and then describing the two colours by examples of things that usually are red or blue, respectively, or by describing the feelings we experience when looking at the two colours. Another approach is to describe the two colours by their wavelength spans. When the length of the light waves is around 650 nm, the colour we see is called "red" and when the length of the light waves is around 450 nm, the colour we see is called "blue". The number, here illustrated by the wavelength, is a way of naming, denoting but not quantifying the colours. Using a quantifiable unit (in this case the amount of millionth millimetres) to identify a quality is thus not the same thing as quantifying the thing that carries that quality.

Another example is the case of carbon dioxide and carbon monoxide mentioned above. To identify the gas, you may let the gas bubble through water and then taste it. If it tastes acidic, we can conclude that it is likely to be carbon dioxide, as this gas will form carbonic acid when dissolved in water, a common ingredient in carbonated drinks. Another way to identify the gas is to use an instrument (called a gas chromatograph) that measures how fast the gas diffuses, since the rate of diffusion depends on the weight of the gas, among other things. Carbon monoxide will travel faster than carbon dioxide because it is less heavy. The time needed for carbon dioxide to move through the instrument is registered and reported by the use of numbers (minutes and seconds), but these numbers cannot be used to quantify the gas, only to distinguish different gases from each other.

If we wish to move from the qualitative, that is identifying the colour or the gas, to quantifying the intensity of the colour red or the flux of carbon dioxide in the surrounding air, we need additional information. The former case, moving from the qualitative to the quantitative, will involve determining the intensity of the light, a rather simple task. When moving from identifying the gas to quantifying the flux from the soil to the atmosphere in a forest, the study becomes considerably more complicated and we have to elaborate our methodology substantially if the results are to be solid enough to serve as a basis for reliable conclusions.

Context dependence and quantification

Moving from a qualitative analysis and description to a quantitative estimate is always easier said than done and requires careful consideration to avoid various pitfalls. In this sense, there is nothing that separates the carbon dioxide example from any other study in which you wish to use a qualitative study as the basis for a quantitative study. To use in-depth interviews with four twelve-year-olds regarding their religious beliefs to draw conclusions regarding twelve-year-olds in general is intricate and extremely difficult, many would argue impossible. To use the results from four air samples that have been carefully analysed for their chemical composition to draw conclusions regarding the composition of air in general is likewise intricate and extremely difficult, many would argue impossible. The former type of studies is seldom done, not even after careful consideration, great hesitation and lengthy discussions. In contrast, it is easy to find examples of the latter type of studies, even in highly respected journals. The only way to understand this discrepancy is that the former group is familiar with general qualitative methods literature whereas the latter group are not (including editors and reviewers).

It is often argued that the exactness of the natural sciences in combination with the fact that the results are globally valid is what makes them more reliable, as they deliver exact answers whereas the social sciences and humanities are always context dependent. This is not the case. The exact sciences (which are only a sub-component of the natural sciences) are often able to deliver exact answers *under ideal conditions*, which actually exist only in theory. With the use of laboratories, it is possible to get close to ideal conditions, and it is possible to achieve almost exact answers, but only rarely is it possible to create conditions which allow for truly exact answers. Outside the laboratory, in more complex environments, the exact sciences can provide only approximations. If exact knowledge is to be useful in a complex context, it must consequently be considered in light of that very context.

Discussions on the context dependence of the natural sciences quite often start (or end) with someone lifting a mug or a box of matches and asking whether or not it may be disputed that the object will fall if dropped. Sadly the core issue is overlooked, since most interesting situations are considerably more complex than a falling mug. Even in such a simplified situation we are only able to draw conclusions only about the fall, and hardly even then: we cannot with certainty know whether or not the mug will break and, if so, if it will break into small or large pieces. We do not actually know if the person in question will let the mug fall, as an act of will is needed to make the fall happen. The only thing we are able to deduce with certainty is the time it will take for the mug to fall, based on an idealized model that assumes free fall without friction, wind, etc. To claim that the exact sciences are context dependent, is in other words to claim that everything must be put into context to have a meaning, which is something completely different from denying basic physical laws under ideal conditions.[8]

By studying processes in the laboratory it is (often) possible to create a (close to) exact understanding of how various physical and chemical phenomena and processes act under ideal or close to ideal conditions; the results can be used to extrapolate to environments outside the laboratory. The laboratory environment makes it possible to minimize and sometimes eliminate natural variation. When conducting an experiment designed to find an exact answer, variation is an unwanted result of the variability of the instrumentation. This stands in stark contrast to environments outside the laboratory where natural variability is an inherent, central and information-carrying feature of the system. Variability is different in different environments and its characteristics hold crucial information. For example, when it comes to biodiversity, an area with one hundred different plant species and ten plants of each species will differ in many aspects from another area, which also holds one hundred different plant species but one thousand plants of one species and only two of the other ninety-nine. An exact answer may thus not be sufficient and may even be misleading when you describe a complex field, as additional information is often needed. To use an oversimplified example: imagine a person holding one hand in a bucket with icy cold water and the other in a bucket with boiling water. It is possible (and easy) to make an exact estimate of the average temperature which is objective, replicable and globally valid. No-one will argue, however, that such an estimate will provide a fair description of the situation the person is experiencing.

Exact information must consequently always be interpreted to be of any use. To interpret any information, however exact, in light of a complex context, is both intricate and complicated. A useful answer concerning a complex question cannot and should not be exact since the exactness will be misleading. I use an example to illustrate that results (the answer to a research question) may be both globally valid and exact or neither globally valid nor exact, depending on the context. I use the following question as my example: "How much will addition of 0.5 ml of nitrous acid decrease the pH of water at a temperature of 20 degrees Celsius?"

Table 1 Illustration of how the results may or may not be exact and globally valid when investigating how much the addition of 0.5 ml of nitrous acid will decrease the pH of water at a temperature of 20 degrees Celsius. The exactness and global validity of the results are context dependent, as they depend upon how "water" is interpreted.

| | Are the results | | |
Sample (Water = ?)	Globally valid?	Exact?	Case
Pure water	Yes	Yes	1
A surface water sample	No	Yes	2
Surface water in a region	No	No	3
Water	Yes	No	4

In the first case in Table 1, where water stands for "pure water", the answer is undeniably both globally valid and exact, provided that the instrumentation and purification methods have been calibrated and the method/instrumentation variation is negligible. Analyses of pure water at any laboratory should give the same result. It is not valid for "water" in general, however, but only for "pure water", in other words distilled, deionized, filtered water, which is only found in laboratories, provided that ideal conditions are prevailing at the laboratory.

In the second case in Table 1, where water stands for "a surface water sample", the variation among replicates of the same sample will be larger than for purified water, since natural water is more heterogeneous than purified water. If a sufficient number of samples are analysed, the average value will give a fair picture of the water from which the sample was drawn – more analyses will not change the results. Hence, providing an exact estimate of the average is sufficient provided that the variation is small. An estimate of the variability should also be provided to make it possible for the outsider to determine whether the variability is relevant or not. The answer can thus be said to be (rather) exact, but it is not globally valid; as discussed in the third case a sample from a different lake or river will give a different result and a sample from the same surface water sampled at another time or at another location may actually also give a different result, since the water quality in the same water body varies with time and space.

In the third case, water denotes "surface water in a region". If we set out to determine the effect on surface waters in a region, we will find that the variation among samples from different surface waters will be considerably larger than the variation among replicates of the same sample due to the heterogeneity of the biophysical world. An exact answer will, in this case, be misleading since the variation among the samples describes the situation and should consequently be included in the answer. *The variation in itself carries information.* The variation is thus not an indicator of accuracy. The answer is not globally valid, since the results depend on which region is sampled.

In the fourth case, water stands for any water sample. If we set out to achieve a picture of how the pH would change after addition of the acid to any water, we will need to collect a large number of samples. As in case three above, we need to include information about the variation as it describes the heterogeneity, i.e. what reality "out there" looks like. And reality in complex fields is seldom linear or normally distributed. Since our aim is to describe the response of any water sample, the sampling should be designed in such a manner that the results are globally valid, but they will certainly not be exact.

In summary: most complex issues dealing with environment, health or security are far more complicated than the water example above. Hence, when the aim is to achieve a deeper understanding of a complex field, it is rarely if ever possible to delimit the study to questions to which there are exact answers. This is for two reasons, as already explained. First, exact answers are valid only under ideal conditions that prevail only in theory and sometimes in the laboratory. To be valid in a more complex context, the answer must be interpreted. For example, the exact effect of nitrous acid added to pure water is globally valid, but only in a laboratory context, since pure water does not exist outside laboratories. Second, (close to) exact answers are possible only when the natural variation has been minimized or eliminated. Environments outside the laboratory are characterized by their heterogeneity and the variation is a central and information carrying feature. Even the most exact findings consequently do not have any meaning in a complex field if they are not put into context.

Interpretation and context

The above serves to illustrate that anyone who studies complex issues is likely to benefit by reflecting upon how the context influences which conclusions may be drawn from a specific study – context is thus relevant not only to students in the humanities and social sciences but to also students in the natural sciences and technology. As mentioned above, there is a rich literature on qualitative methods, which commonly discusses challenges emanating from the context. Many of the questions that are discussed in this literature are of a general character and even though not spelled out, many of the issues discussed are general and highly relevant across The Gap. There is yet another body of literature that describes and discusses the production of knowledge in science.[9] This literature is also written for (and mostly read by) students who study sociology, philosophy and history of science, and not for students studying science.

It is sometimes argued that the natural sciences and engineering differ from the social sciences and humanities in that the former see the whole as the sum of the parts whereas the latter emphasize the need to study the parts and the whole interchangeably. In the philosophy of science literature, the former approach is often called reductionist whereas the latter is a prominent part of a hermeneutic approach. The latter approach builds upon the idea that the whole is more than the sum of the parts whereas the former builds upon the idea that the whole can be understood as the sum of the parts. Hermeneutics was originally a method that dealt with the interpretation of texts, mainly religious, but it has later come to denote an approach rather than a method. In simplified terms, a hermeneutic approach means that you believe that nothing can be understood without interpretation and that it is therefore necessary to remain open to other potential ways of interpreting the phenomenon under study. Furthermore, the understanding that the parts and the whole are interchangeably dependent on each other makes up a crucial part of the interpretation process. Since hermeneutics to a large extent deals with the interpretation of texts, it comes as no surprise that computer communication, especially artificial design, draws upon such knowledge.[10]

The importance of studying the parts and the whole interchangeably, as well as the need to understand that data must be interpreted to be given meaning (also called a holistic approach) are applicable to all research that deals with complex issues. There are numerous examples of research traditions within the natural sciences that emphasize the need to focus on the parts and the whole interchangeably to understand an issue, since it is impossible to understand the whole as simply the sum of the parts, especially in the field-based traditions mentioned above. Basic textbooks in disciplines (i.e. hydrology, hydrochemistry, ecosystem ecology, limnology and eco-toxicology) generally emphasize the necessity to study the system and not only the parts, since the sum of the parts, due to the numerous feedback loops and related complexities, will not add up to the whole. It is stressed that the information is context dependent, but the literature rarely if ever discusses the challenges of context dependence on a general level and it seldom draws upon the rich method literature from the humanities and social sciences.

Notes

1   Charles Percy Snow 1993.
2   John de la Mothe 1992.
3   See, for example, *The SAGE Handbook of Qualitative Research* by Norman K. Denzin and Yvonna S. Lincoln (eds) 2005.
4   In order for a study to be called qualitative in the humanities and social sciences, many demand that the study in one way or another contains a reflection on the meaning of what has been studied, a demand that is rarely or ever fulfilled in qualitative natural science studies.
5   See, for example, Roger Strand 2002.
6   See, for example, Daniel Sarewitz *et al.* 2002; Lorraine Daston and Peter Galison 2007; Lynn White Jr 1974
7   See, for example, articles in the journal *Science, Technology and Human Values.*
8   For a discussion on the fact that all knowledge is and must be contextualized, see Ludwik Fleck's *Development and Genesis of a Scientific Fact*, originally published in 1936, reprinted in 1979.
9   See, for example, Ludwik Fleck 1979; Thomas Kuhn 1970; Bruno Latour 1987 and Donna Haraway 2008, to mention a few.
10 John C. Mallery *et al.* 1986.

3 Questioning to learn and learning to question

Interdisciplinary work may be conducted in a group or by a single person. In either case, you will encounter diverging expectations from the disciplines you use and the scholars with whom you interact. Sooner or later, you will face the need to reconcile diverging expectations from the disciplines you draw upon and from those you interact with. The aim of the framework presented in this chapter is to help you draw upon the reconciliation process and use it as a tool in the creation of exciting, rigorous and cutting-edge work. By asking the questions in the framework you will be empowered to gauge the quality of your work and it will help you clarify the quality of your work to others.

The increasing demand for interdisciplinary education and research stresses the need for a definition of credibility that allows for variations in form without compromising rigour. In other words, students who engage in endeavours involving more than one discipline need to be aware of the fact that good research comes in many forms and, not least, that teachers and supervisors might not be aware of this fact. It is easier said than done to broaden your understanding of quality, but certainly more likely to happen if you strive to uncover, discuss and justify the explicit as well as the implicit norms guiding various perceptions of good vs bad studies. The intention of the framework presented in this book is to facilitate such dialogues.

The first part of the framework deals with assessment of the "inter" in interdisciplinary work (Chapter 4). This will help you clarify why you conduct interdisciplinary work in the first place and it will help you analyse whether the way you go beyond traditional academic cultures strengthens your study in relation to its aim. The second part of this framework deals with your ambition in relation to the surrounding society or, in other words, whether you wish to interact with the society outside academia and why (Chapter 5). The third part of this framework focuses on the academic

Interdisciplinary Environmental Studies: A Primer, 1st edition. © Gunilla Öberg.
Published 2011 by Blackwell Publishing Ltd.

rigour of your work (Chapters 6–11). It will empower you to conduct work of high scholarly quality and it will help you clarify that scholarly quality to others.

Part I: Interdisciplinary expectations (Questions 1 to 3)

Interdisciplinary studies can be conducted for very different reasons and to very different ends.[1] As further outlined in Chapter 4, your work will be unnecessarily difficult if you, your supervisor and your committee members have different understanding of why you conduct interdisciplinary work. What are your expectations? What do you hope to achieve by using an interdisciplinary approach? As further discussed in Chapter 4, it often helps to start by gauging where you are on the reflection scale (Question 1). The scale is intended to force you to defend and clarify your position to yourself and to others. As a second step, it often helps to clarify your aspirations (Question 2), which means to clarify why you are conducting interdisciplinary work in the first place. Wait with Question 3 until you are about halfway into your work.[2] At that stage, it is usually a good thing to take a bird's eye perspective and analyse your work as an outsider. But try to avoid asking this question too early. I explain in Chapter 4 why I believe it is wise to wait.

Question 1:   Where do you position yourself on the reflection scale?

Question 2:   To what end are you using knowledge from different disciplines? Is your ambition to:

    a)   solve a practical problem?[3] (Pragmatic).

    b)   contribute to a new or emerging field? (Conceptual).

    c)   describe/analyse/understand in what way disciplinary structures cause problems? (Epistemological).

    d)   describe/analyse/understand in what way societal perceptions of the issue at hand cause problems? (Ontological).

Question 3:

    a)   In what way(s) does your way of using knowledge from different disciplines leverage understanding in relation to the aim?

    b)   What would be lost if one or several elements was excluded?

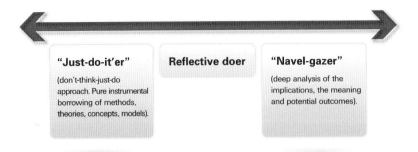

Figure 5 An illustration of a reflection scale with the (consciously exaggerated) end-points being those who just do without thinking (of which engineers often are accused) and those who just think and never do (often pinned to the humanities). Discussing the scale and identifying where you position yourself is likely to help you avoid unnecessary misunderstandings with your supervisor and committee members.

Part II: Transacademic aspirations (Questions 4 and 5)

Why do you wish to interact with the society outside academia? As discussed in Chapter 5, one of the key obstacles to transacademic work is that people interact with others for a variety of reasons while taking for granted that the interaction is driven by common interests. By assuming the opposite, that those involved in the project have *differing* understanding of the reasons to interact with society, you will not only be able to avoid some difficult pitfalls, you will be able to use this fact to help you conduct exciting research. Make a point of, from time to time throughout the work, raising Questions 4 and 5 with your supervisor, committee members and other people you collaborate with. It helps to be specific but brief. Outline what your aspirations are and what you hope to achieve by interacting with the extra-academic world. Scrutinize why you interact with society, in which steps of the research process you interact, with whom you interact, and how this strengthens your study. These questions are designed to facilitate the process of clarifying your transacademic aspirations:

Question 4:   Who participates in which parts of the study and how?

Question 5:

        a)   How does this (Q4) bring leverage to the study in relation to its aim? (Be specific)

        b)   What would be lost if something were omitted? (Be specific)

Part III: Academic rigour (Questions 6 to 10)

An academic study must be rigorous from an academic perspective. As further discussed in Chapter 6, one of the major challenges for interdisciplinary work is that "a good paper" looks very different in different disciplines. The third part of the framework presented in this book is built around five core areas in which all empirical disciplines share common ground but the commonalities are hidden by the large differences in form and style. In this part of this framework, the questions deliberately include terms such as "sufficiently", "coherently" and "reliable", which are unproblematic in a group with shared norms but become increasingly ambiguous as diversity increases. Pondering these questions, alone or in a group, will lead to increased awareness of your own perspective, and facilitate dialogue, collaboration, and thus creation of common ground. It will help you conduct more rigorous work and it will also help you clarify the quality of your work to yourself and others. Each of the following questions are discussed in Chapters 7 to 11.

Question 6:   Is your study sufficiently and coherently demarcated?

As further outlined in Chapter 7, all scholarly work must be demarcated, as it is impossible to study everything. You must decide what to study, from what perspective, what information to gather, and how and where to gather that information. A good study clarifies the grounds upon which the choices are based. The question will likely inspire dialogues on how to demarcate a study, which choices should be made visible and the pros and cons of various ways to clarify the choices made.

Question 7:   Has the information been collected in a reliable manner and is it of sufficient quality?

As further outlined in Chapter 8, a reliable empirically based study demands an ability to collect researchable information. You must, in other words, be able to figure out what type of information or evidence to collect, which will bring light upon the issue in focus and how to collect the material so that it holds for scrutiny. All procedures demand craftsmanship as well as an ability to critically evaluate the limitations of the used technique. The question is likely to inspire dialogues of what type of conclusions you may draw from what type of material – dialogues which will be most useful in the formulation of your studies.

Question 8:   Is the study material sufficiently anchored in relevant literature in terms of the framing, methodology and analysis?

As further outlined in Chapter 9, anchoring the work in relevant literature is a cornerstone of all scholarly work. Authors must be able to distinguish between citations, accounts, plagiarism and their own conclusions and must clearly spell out from where statements, comments, facts and conclusions originate. It is of the utmost importance that it is clarified which statements result from the study and which statements originate from someone else. The anchoring must be clear all the way through the framing of the study, the method and the analysis. The question is likely to inspire a dialogue of how to interpret the words "sufficient" and "relevant", a most appropriate dialogue that ought to be kept alive and continuously carried on in all academic environments.

Question 9:   Is the information analysed in an informed and reflective way?

Anchoring the work in relevant literature is a necessary but not sufficient requirement for an academic paper to be assessed as credible. As further outlined in Chapter 10, a common demand in academia is that all information used must be presented and viewed with a critical eye. Veronica Boix Mansilla suggests focusing on "the capacity to use knowledge over that of having or accumulating it."[4] Question 9 is likely to inspire a dialogue regarding which information you are expected to manage critically and which information you might treat as common knowledge, or in the words of Bruno Latour,[5] what kind of information we accept as "ready made science" and what we treat as "science in the making." Surely, it is impossible to always question everything, but an ongoing dialogue on how we do distinguish the former from the latter is likely to nurture a lively and fruitful environment.

Question 10: Are the form and structure in line with agreed norms and does the text consequently and consistently follow the chosen form and structure?

To place a study in context by relating it to relevant sources and clearly distinguishing between your own conclusions, thoughts, ideas and results and those of others is in part a question of content and in part a question of form, and form and content are intricately intertwined. Chapter 11 discusses the fact that texts written by students in interdisciplinary contexts often take a shape that is rather unfamiliar for most of the involved teachers and researchers and even less familiar for persons outside the group. In order to gain recognition outside one's own environment, it becomes even more crucial than in traditional academic environments to make conscious choices not only regarding aim, theory and method, but also regarding form and structure. To be able to choose, you must be aware of the options and you must acquire sufficient skills to manage the chosen style. The question is likely to inspire dialogues regarding how to interpret "agreed norms", which is a discussion that far too seldom is carried out.

Notes

1   See, for example, Julie Thompson Klein 1990; Liora Salter and Alison Hearn 1996; Sharon J. Derry *et al.* 2005; Andrew Barry *et al.* 2008.
2   In Chapter 4, I explain in more detail why I believe it is wise to wait with Q3.
3   Note that "problem" here refers to academic problems. In other words, the problem is that you cannot answer your research questions unless you go outside the borders of traditional disciplines. To solve this problem you borrow methods, theories, models, concepts and thought styles from various disciplines. Issues related to "real world" problem solving are dealt with in the second part of the framework (Q4–5).
4   Veronica Boix Mansilla 2004, p. 4.
5   Bruno Latour 1988.

4 Why do you conduct interdisciplinary work?

Always remember: your research questions should guide your study. When you run into people who question your credibility, it is easy to forget that your intention is to become an expert with respect to the aim of your study – not to become an expert in a number of disciplines. If instead of focusing on your research questions you try to answer the question: "What is it that makes this study interdisciplinary?" your energy is likely to be focused on identifying key disciplines of your work. To become "sufficiently" knowledgeable in a discipline is in itself immensely time consuming, since disciplines have fuzzy borders; it can always be debated if you know enough to count as an expert. You risk finding yourself reading to become "sufficiently" knowledgeable in each of the identified disciplines. And you will always find yourself insufficiently knowledgeable. No matter what, you will *always* encounter people who will argue that you do not know enough in discipline X.

There is a way to avoid this trap: do your best to figure out how the aim and your research questions are understood by different parties. Whom have you invited to discuss your research questions and how will that influence the way you frame the questions? What literature can help you understand the issue in different ways? What methods can provide you with material that enables a deeper understanding of the issue at hand? What skills do you need to acquire to carry out the studies? If you problematize the perspectives that guide your understanding of the issues at hand and pay serious attention to the problem formulation process, the study will more or less automatically become interdisciplinary.

Where do you position yourself on the reflection scale? (Question 1)

The level of reflection is often a point of disagreement in an interdisciplinary project. On one end of the spectrum are those who reason as follows:

Interdisciplinary Environmental Studies: A Primer, 1st edition. © Gunilla Öberg.
Published 2011 by Blackwell Publishing Ltd.

"I formulate the aim of my study, I then look around and pick the theories, methods, concepts, models and whatever I need to answer the questions of my study. I stay clear of disciplinary snobs and do my thing. I solve the problem, I find a good place to publish the work and that's it. No need for reflection – that only slows down the process." This approach is often driven by empirical considerations, while theoretical considerations are secondary. Interdisciplinary work is mainly an act of practical borrowing (Figure 6).

At the other end of the spectrum we find those who claim that unless *all* theoretical issues are solved, a project should not be carried out. They reason as follows: "It is well known that knowledge cannot simply be moved from one context to another without being transformed. To say that it is possible to 'borrow' something from a certain discipline is naïve, to say the least. It is necessary to have a deep understanding of the disciplinary context before anything can be sensibly used in a new context. One must analyse what it *means*. In addition, one must analyse the implications as well as potential outcomes." This approach is often driven by theoretical considerations, while empirical, technical and practical considerations are secondary.

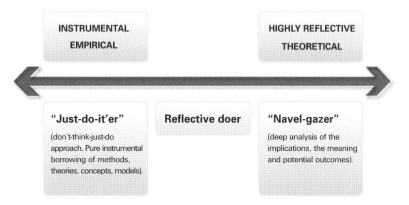

Figure 6 An illustration of a reflection scale with the (consciously exaggerated) end-points being those who just do without thinking (of which engineers often are accused) and those who just think and never do (often pinned to the humanities). Discussing the scale and identifying where you position yourself is likely to help you avoid unnecessary misunderstandings with your supervisor and committee members.

Both a purely instrumental approach and a highly reflective approach are acceptable in certain mono-disciplinary and multi-disciplinary settings.

There are lots of examples of people digging deeper in the bottom of a pit that others have dug. A purely instrumental approach can, for example, apply a certain model to a new geographical area. This will not produce a highly original study but it will contribute to the field. A highly reflective thesis can, for example, dive deeper into the meaning of a concept that is of crucial importance to the field.

Even though both the instrumental and the highly reflective approaches may be acceptable in mono-disciplinary and multi-disciplinary settings, they become problematic in interdisciplinary settings. Without conscious reflection, you are at great risk of unconsciously being guided by the tacit norms of the disciplines you use. Pretending that disciplinary boundaries do not exist when you are looking for theories, methods and other research components will not make the boundaries go away. In fact, there is a great risk that by pretending the boundaries do not exist, you will become their prisoner. It is only by being aware of the differences and consciously chal-lenging the tacit norms that you will be empowered to go beyond the traditional. By being aware of the disciplinary differences and boundaries, you will gain power to use them or ignore them, whichever suits your needs. Being aware of and thereby able to use disciplinary differences is one of the great opportunities of interdisciplinary work. The other end of the spectrum is problematic in a different way. By moving to the outer end of the scale the study is transformed to a meta-study (a study of a study). This is what scholars in philosophy or sociology of science do. Such studies are immensely valuable but they are hardly interdisciplinary; there are disci-plines that are able to provide methods, theories, concepts and modes of thought to shed light on these issues. Combining reflective elements with other aspects however, makes a study interdisciplinary (i.e. moving from the outer right end of the scale). The most fruitful interdisciplinary projects are those that manage to strike a balance between "just do it" and "oh, we have to do some more thinking." In other words, if you have the ambition to become an interdisciplinary scholar, you should strive to become a "reflective doer".

In the course of writing this book, I have invited students and colleagues to mini-workshops and seminars to discuss various sections of the manu-script. I wish to emphasize that this book could not have been written without these contributions. As part of this interactive process, I asked four recent graduate students to reflect upon their studies in relation to Questions 1–5 to be used as illustrations in this book. Below follows short descriptions of their theses to help you put the reflections into context. Thereafter follows their reflections in relation to Question 1. Further on in the text, you will find their reflections on the other questions.

Example 2
Shannon Hagermann's thesis *Adapting conservation policy to the impacts of climate change: an integrated examination of ecological and social dimensions of* change:[1]

In my thesis, I examined the challenge of designing and implementing new goals and management strategies for biodiversity policy tailored to the impacts of climate change. Using a mixed methods approach, the results reveal that many experts see the need for transformative changes in conservation policy that include new strategies like facilitating species distributions through disturbance, and revised goals and standards of success. Yet simultaneously there exists public precautionary ambivalence and resistance to anticipated difficult trade-offs implicit in emerging alternatives. I argue that prevailing commitments to more preservationist (less ecosystem engineering) values have in part shaped the adaptive response so far. Combined, the results highlight that policy adaptation to climate change impacts within "science-based" conservation is a tangle of technical understanding and social dynamics.

Shannon Hagermann, October 2009

Reflections on Question 1: Overall, I would say that I strive for the middle of this scale, but I don't feel anchored there. My experience has been that I toggle back and forth. For example, over the course of my PhD studies, I was at times a navel-gazer (mostly the beginning while conceptualizing the work, and at the end in interpreting the meaning of it). While at other times (the middle/research design), when I was focused primarily on implementation, I was a just-do-it-er. When I began researching the challenge of adapting biodiversity policy to the impacts of climate change and other drivers, I first needed to understand the different sources of knowledge and theoretical perspectives that were potentially relevant. In part, this involved understanding knowledge from the systems ecology, complex adapting systems and climate change impacts literature. However, if I had taken a strictly just-do-it approach I would have failed to reflect on the sometimes "taken for granted" implications of the meaning of knowledge produced within these disciplines and with respect to my topic. Some systematic navel-gazing was essential because it was critical to understand how meaning through science has been attached to concepts such as nature, biodiversity and wilderness, and the material implications that these various conceptions have had for policy objectives.

When it came to doing the empirical research (and borrowing methods and relevant theories that I was now familiar with) I was acting in the sense of "just-do-it", but with an awareness of the importance of always reflecting on the meaning of my data and how meaning is shaped through my chosen methods, questions and conceptual frame.

So I might say that I engage with both ends of the spectrum as a function of research trajectory. Maybe being a reflective doer means you know where along the scale you are at various points in your research. With apologies for gross caricature, my intention in practising interdisciplinary research is not to be immobilized at the right-hand end of the scale, nor cavalier and unreflective as might characterize the left, but somewhere in the spirit of a reflective do-er.

Shannon Hagermann, October 2009

Example 3
Natalie Ban's thesis *Multiple perspectives for envisioning marine protected areas*:[2]

My thesis provides the first direct comparison between – and integration of – community-based and science-based approaches to the establishment of marine protected areas. I explored and analysed these approaches, separately and together, in two areas in British Columbia, Canada. First, I generated a community-based plan for marine protected area placement through partnerships with two First Nations (indigenous peoples) in British Columbia. Second, I applied a conservation planning decision support tool (Marxan) to determine marine protected area placement under scientific precepts. Third, I integrated community-based and science-based approaches, to find that they verified and complemented each other. My research shows that community involvement in placing marine protected areas can help meet many ecological goals, although biophysical data improve the conservation value of sitings.

Natalie Ban, October 2009

Reflections on Question 1: I consider myself a "reflective doer". During my PhD, I explored various approaches I could apply to the questions I was addressing. Rather than simply borrowing approaches, though, I was interested in fully understanding – and hence being able to justify my use of – such approaches. In interdisciplinary work in particular, there are likely several ways of answering a research question. A crucial challenge for the researcher is to fully justify the approach selected. Simply borrowing, without contextualization, is bound to be criticised. On the other hand, the need to explore multiple approaches to select the most appropriate one for me meant that reflection was necessary, but I had little time to do "navel-gazing". Pure reflection without doing would not have achieved my aims.

Natalie Ban, October 2009

Example 4

Charlie Wilson's thesis *Understanding and influencing energy efficient renovation decisions*:[3]

My thesis was motivated by the enduring policy problem of how to improve the energy efficiency of existing homes. It's an enduring problem because policy-makers have limited recourse to coercive measures. Our homes are our proverbial castles: policies in pursuit of social objectives can, at best, only hope to influence our choices and behaviours at home. My interest was in energy efficient renovations. I reasoned that to understand how best to influence homeowners' renovation decisions, I needed to understand those decisions. I reviewed decision models from various disciplines and used them to design a survey for testing a host of hypotheses about decision-making. The survey went out to around 800 homeowners who were either contemplating, midway through, or had recently completed a home renovation. Comparing these cross-sections allowed me to see how the determinants and structure of the decision changed through the process. This revealed a tendency for homeowners to rationalise their decision after the fact. A renovation that may have originally been motivated by emotions or social norms was later justified by environmental and financial reasoning. This finding is important as it suggests prevailing policy models are misplaced in their emphasis of financial influences (targeted by incentives, subsidies, grants) and environmental influences (targeted by education, information and exhortation). Using amenity contractors to bundle low-cost efficiency measures up with emotionally-charged decisions to have kitchens and bathrooms remodelled may be a simpler and more effective way to go.

Charlie Wilson, October 2008

Reflections on Question 1: I've moved from left-to-right on the reflection scale as my PhD research has progressed. When I started out, I was genuinely perplexed at the incredible differences between how different disciplines answered the question I was interested in: why individuals behave in the way they do.

It seemed obvious to me that I could just compare how well each disciplinary model fared in explaining behaviour in a particular context.

I could then integrate the bits that worked into a new and improved interdisciplinary model. So I just started raiding different research traditions, finding their most successful or prominent models of behaviour, and developing hypotheses that would allow me to test these models.

However, as the research progressed and became all confused by the data, I began to see that the models differed for good reasons. They were answering questions about behaviour at different scales, of different types, subject to different influences, and so on. I became more and more caught up in the task of integrating these models without producing something so specific to the empirical context I was looking at that it would lack any generalisable applicability. I had moved from just-doing-it to navel-gazing, and to complete the research, I had to compromise more, I hope as a reflective doer.

Charlie Wilson, October 2008

Example 5
Jane Lister's thesis *Co-regulating corporate social responsibility: government response to forest certification in Canada, the United States and Sweden:*[4]

Through a comparative case study that drew on more than 120 interviews across 3 countries, my dissertation evaluated how and why governments within the world's leading certified nations responded to forest certification, and the implications for forest governance. I showed how the Canadian, US and Swedish governments are increasingly engaging in certification through a range of co-regulatory approaches that supplement rather than substitute for forest laws. The cases also highlighted how certification co-regulation benefits forest administration and policy-making suggesting that governments are engaging in certification for other than market-driven reasons. My dissertation therefore challenged the theory of "non-state market-driven" governance, arguing that certification is more accurately classified as a co-regulatory mechanism. I developed the new concept of CSR co-regulation and introduced three analytical tools to evaluate the co-regulatory arrangements and establish a framework to facilitate future research in this area. Beyond the theoretical contribution, the findings also offered practical guidance to policy-makers seeking to understand forest certification and adopt new adaptive forest governance approaches.

Jane Lister, November 2009

Reflections on Question 1: I was an intense naval-gazer at the very start of my PhD program and then was pretty much a "just-do-it-er" (extremely busy just trying to keep my head above water). My decision to leave my professional environmental consulting career to do a PhD was very much tied to in-depth disciplinary and methodological reflection. I knew I wanted to research forest governance but could I find a suitable academic home to pursue my studies? I quickly realized as I interviewed professors across the various departments that there was quite a difference in the questions they asked, their assumptions, and their research methods. I remember being told such things as, "unless you're prepared to do a statistical survey I'm not interested." I ended up in an interdisciplinary department and chose my PhD supervisor and committee very carefully, based not only on their area of expertise (within their disciplinary departments) but also their methodological approaches and openness to interdisciplinarity. As I heard stories over the next four years of students in agonizing struggles with their committees, I realized my early reflection was an extremely worthwhile investment.

Jane Lister, November 2009

To what end are you using knowledge from different disciplines? (Question 2)

If interdisciplinarity is a means to an end, the intention of Question 2 is to help you envisage that "end". The core question is what you intend to achieve by using an interdisciplinary approach. It is clear that different people have different ends in mind. The challenge is that most assume a shared understanding of the end, even though the end seldom is explicit. Drawing on the literature,[5] I have identified four different ends that are common in environments where people deal with the interaction between human and natural systems: Pragmatic, Conceptual, Epistemological and Ontological. These four ends are not only common – they are also a common source of confusion. Sorting out which one(s) of the four ends you strive for will help you avoid unnecessary hurdles, and it will help you focus your study.

To illustrate what I mean by "ends", I start with a reflection on my own thesis followed by reflections by the four graduate students introduced above. I used a number of natural science disciplines, Shannon and Natalie draw on knowledge from the social sciences and humanities as well as natural sciences, and Charlie and Jane use a number of social science disciplines.

Example 6
My own doctoral thesis *On the origin of organohalogens found in the environment*

In the 1980s, it was generally agreed that organo-chlorine compounds did not occur naturally. Based on this assumption, a method which measured the total amount of organo-chlorine (AOX) in water was launched as a tool to monitor human impact on surface water. A group of scholars were studying the effect of pulp mills on the quality of surface water and among other things they measured AOX. To their surprise they found high concentrations in remote, uninhabited areas. I was invited to join the group and asked to answer the question: What if this isn't pollution, what if it is something natural? By combining theories, methods, techniques, models and concepts from environmental chemistry with (soil) microbiology, taxonomy, pharmacology, enzymology, hydrochemistry and hydrology, I came to the conclusion that organically bound chlorine was produced naturally in soil. It was very much a question of borrowing various elements from different disciplines and combining them in a new way.

Pragmatic approach: I need to solve a problem

In hindsight, it is quite obvious that I used a rather pragmatic approach. The group wanted me to solve a problem and answer a specific question (is the AOX in surface water of natural origin?). This illustrates the most common approach to interdisciplinary work; that of borrowing something (method, concept, theory, model, thought style) from another discipline or field. The discipline in which one is rooted does not have the right tools and therefore one seeks help outside one's own box. Or the way one has formulated the aim does not fit in any particular discipline, so one borrows from a number of different disciplines. The focus is often to solve a methodological problem.

The literature on interdisciplinary work speaks a great deal about the dangers of an instrumental approach and the risks of not understanding the consequences of the studies one conducts. There seems to be confusion between an instrumental (non-reflective) and a pragmatic approach. You can choose a pragmatic approach without any reflection at all but it is also possible to be highly reflective and still choose a pragmatic approach. There are numerous examples of fruitful reflective studies that are based on pragmatic borrowing, and I would say that it is pure snobbishness to claim that the other three approaches outlined below are better or "more" interdisciplinary. They are simply different.

Conceptual approach: I want to contribute to a new or emerging field

Going back to my own work: in the iterative work of reading literature, designing our empirical data collection procedures, collecting and analysing the data and honing and rephrasing the aim, "chlorine biogeochemistry" or "the chlorine cycle" started to emerge as a new area of inquiry. The work went from being purely pragmatic to being conceptual. The aim was extended from answering if the AOX was natural or not to asking why organisms produce organic chlorine, how they do it and under what conditions. Something new was emerging. When have you created "deeper" knowledge in a given discipline or field and when have you created knowledge in an emerging field? There is no clear distinction, but at a certain stage, you have created something that emerges outside the traditions you started with. You have embarked on a journey into an emerging field. This can be called a conceptual approach and one sign that this is happening is the introduction of new concepts. To some extent it is a question of integrating existing elements into a given framework, to synthesize existing knowledge, and to some extent it is a question of creating something new.

Epistemological approach: I find disciplinary way(s) to describe the issue problematic

During my doctoral studies, it became increasingly clear that the limitations created by disciplinary borders was a severe obstacle for my field of study and I wanted to introduce a chapter in my thesis discussing these issues. My supervisor firmly advised against this and I eventually gave up the idea. As a post-doc, I wrote a paper in which I discuss the problems caused by disciplinary borders with regards to our understanding of the chlorine cycle.[6] The discussion deals with problems related to different ways to structure knowledge, and differences among disciplines in what they consider to be of value, which could have been the focus of a thesis driven by epistemological aspirations.

The difficulties involved in the creation of new or emerging fields often lead to critique of disciplinary ways of describing or understanding the problem. It is less a question of integrating knowledge from different academic cultures and more a question of analysing the problems that arise from the way the problem is studied, understood and described by the disciplines. With an epistemological approach, the focus of your study is the knowledge itself, how it is structured and how this way of structuring the issue is problematic.

Ontological approach: I find the way society perceives the issue problematic

The reflections and inquiries that were born through dialogues with colleagues in the department led me as a doctoral student to think about how the way society perceives an issue determines whether it is considered a problem or not. With a different academic training (and other sources of funding), I could have written a thesis that focused on how the perception of what is "natural" influences the way organic pollutants are managed. In that case, one could say that I had chosen an interdisciplinary perspective for ontological reasons.

Analysing in what way knowledge is structured and why this may be problematic is closely tied to the fact that the ways we describe things guide our understanding and perception of the world. The aim of your study may be to show that the way we describe and thus understand an issue is problematic. This can be described as an ontological approach. Rather than the knowledge itself, the focus of your study is our understanding of the world.

Reflections in relation to Question 2

I engaged in interdisciplinary work for my doctoral research for both pragmatic and conceptual reasons. When I began my studies, existing proposals to address the challenge of how to adapt conservation policy to the impacts of climate change were largely informed by insights from the natural sciences (within the disciplines of ecology and biogeography more specifically). But conservation decisions are shaped by human preferences in specific contexts, and with impacts for different groups across different scales. Thus my motivation for using knowledge from different disciplines was pragmatic in the sense that by synthesizing and integrating insights from across the natural and social sciences, I sought to provide practical insights for decision-making. My motivation was also conceptual in that one of my research aims was to bring together previous disparate insights under a common integrative framework as applied to this particular problem domain.

Shannon Hagerman, September 2009

—

The overall aim of my doctoral thesis was to address a real-life issue. I investigated a multi-faceted and complex problem – protecting the ocean – from several angles, using interdisciplinary techniques. More specifically, the main question I asked was: How can we select areas for protection in the marine environment that are socially acceptable and ecologically viable? Social acceptability is crucial for having effective protected areas that are supported by communities who comply by the park's regulations. Ecological viability and representation is essential for the persistence of biodiversity. My thesis therefore addresses an issue of current policy relevance given global biodiversity declines, the increasing human dominance of coastal ecosystems, and the commitments of many nations to establish marine protected areas. The main question of my thesis also has academic and theoretical relevance, as it addresses an ongoing debate in the literature about the advantages and disadvantages of different prioritization approaches.

I did not set out to produce an interdisciplinary thesis for the sake of being interdisciplinary. Instead, I intended on solving a problem, and comparing and integrating different approaches to that problem. My thesis can therefore be categorized as taking a "pragmatic approach" and "conceptual approach".

I applied two common approaches to the problem of marine protected area selection: community-based selection, whereby people within my study areas decided on areas preferred for protection; and a science-based approach, whereby the best available scientific data (e.g. species distributions, habitat types) informed potential sites for protection. These approaches have origins in different disciplines. Community-based selection draws primarily from anthropology, using interviews to elicit people's preferences and opinions. Science-based selection originates in biology and ecology, and uses geographic information systems techniques and site selection algorithms. Applying these two approaches firmly places my thesis into a "pragmatic approach".

The goal of my thesis was not only to apply two approaches to prioritizing areas for protection, but rather to integrate these approaches – this places my thesis into the "conceptual approach". It was apparent from the literature that one approach by itself would not provide innovative answers. Through the integration of a community-based and science-based approach, I created a new method for identifying priority areas – a method that integrates community preferences and scientific conservation objectives.

Both approaches I used and integrated could be considered part of the field of conservation biology. Conservation biology draws from many other disciplines, and some now consider it a discipline. Does that mean my research was not interdisciplinary? It is easy to get lost in the semantics of what defines a discipline. Ultimately, I drew upon techniques which require vastly different skills – interviews with people, and computer simulations with ecological data – and therefore in my mind I characterize my research as interdisciplinary.

Natalie Ban, May 2009

—

My research was motivated by purely practical reasons. There has been an ever-growing recognition that our behaviour as individuals contributes to environmental problems. At the same time, there has been an ever-growing use of "soft" policy instruments by governments that seek to influence rather than regulate our behaviour. Surely, I thought, if we understood why individuals behave in the way they do, then we can design policies that effectively influence that behaviour.

I decided to look specifically at individual or household decisions about energy efficiency, an important current policy topic. I scoured different literatures to see how their models and theories could help identify ways to influence these decisions. I began in an entirely pragmatic way to look at integrating the best bits of the different models.

Being in an interdisciplinary research programme, and having a wide-ranging set of interests, I cast my net widely, right across the behavioural sciences. I built up a sort of compound view of this knotty problem: how and why people make decisions about energy efficiency in the home. But as this insect's eye grew, and I became more familiar with different, and let's face it, competing disciplinary approaches, I realised more and more how disciplines would say very similar things in wildly different ways. An economist might talk of the long-run price elasticity of demand, whereas a sociologist might talk of the structuring of choice by socio-technical systems.[7] My eureka moments were realising they were saying the same thing but using the language and methods of their own disciplines. So I became more interested in the epistemological dimension to what I was doing, at the obvious expense of the more pragmatic task in hand.

Charlie Wilson, December 2008

—

My research was driven by pragmatic idealism, conceptual ambition, ontological concerns, and epistemological confusion. I wanted to help solve the problem of achieving more effective forest governance. I had an ambition to understand and contribute to the theory of new tools of environmental governance. I was concerned with the way the academic tradition understood and defined governance authority to reside solely within the state. And finally, I was confused by the way society assumed a black-and-white distinction between public versus private policy approaches when the empirical reality was a blurring and overlapping of the boundaries between the market and the state.

Jane Lister, November 2009

What makes your work interdisciplinary? (Question 3)

When your study is starting to take shape and you start to analyse, synthesize and put the pieces of the puzzle together, I would say it is time to take a bird's eye perspective and ask the question: What is it that makes my work interdisciplinary? A key reason to clarify what the "inter" of your work is that when you conduct work that goes beyond traditional borders, you open up for a larger and more heterogeneous audience than traditional, disciplinary work does. Opening up to a larger audience inevitably means that you also open up for more people who may (and will) criticize your work. This is a good thing, because it means more people will read your work, and if they criticize it, you have managed to make them reflect upon what you say. It has touched them to such an extent that they want to comment, to criticize. Recognize that this is a gift (even though the gift-wrapping sometimes may be a bit whacky). A lot of research passes unnoticed, so rejoice! You have managed to carry out work that people care about. When criticized, you must be able to explain and clarify (rather than defend) not only the quality of your work but also its originality.

A second reason to scrutinize the "inter" of your work is that it will help you identify the academic traditions you draw upon, which in turn will help you to choose a style and a format that is acceptable in those disciplines. It will also help you identify when you break style and format norms. This will make it possible for you to either choose to revise your text in accordance with the norms or expand your explanations to clarify why you do things the way you do. A third reason to scrutinize the "inter" is that it will help you revisit your aim: What is it that you want to achieve with your study and are you consistent in what you do? It is important to realize that you will not be able to cover as much ground as you originally set out to do. This is unfortunately the name of the game. Accepting this fact makes most studies less frustrating. Conducting a study in an emerging field makes it more difficult to see where the study fits as compared to a disciplinary study that digs deeper into a given field. The nature of an emerging field is its unknown character, the vast landscape of the undiscovered. A contribution in a dark seemingly endless space may seem both miniscule and humongous compared to a similar contribution in a fairly well-known field. With no or few points of reference it is hard to know when a contribution is "sufficient". It is therefore generally harder for those who work in an interdisciplinary environment to assess if their contribution is sufficient, as it often seems so excruciatingly small in light of initial intentions and at the same time impressively large in relation to the nothingness that was there before. As a result, it is consequently tempting to cover too much ground and thus risk the quality of the work. Taking a bird's eye perspective and revisiting the aim in light of your interdisciplinary aspiration is usually helpful. A fourth reason to scrutinize the "inter" is that it will help you explain your interdisciplinary aspiration to your supervisor, your committee members and examiners.

The third question of the framework is designed to help you uncover the things that make your work interdisciplinary and it is easier to answer the question if it is divided into the following two sub-questions:

a) In what way(s) does your way of using knowledge from different disciplines strengthen your study?

b) What would be lost if one or several elements was excluded?

The fact that your work in one way or another rests upon the use of knowledge from different disciplines is what makes your thesis interdisciplinary. This should not be confused with your position on the reflective scale (Figure 6, p. 31), or the reasons you conduct interdisciplinary studies (pragmatic, conceptual, epistemological or ontological). Identifying *what* it is that you use, combine and integrate from different disciplines, in concrete terms, will help you clarify the "inter" in your thesis.

To illustrate how Question 3 may be used, I start with a reflection on my own work, which is followed by reflections by Shannon Hagerman, Natalie Ban, Charlie Wilson and Jane Lister.

Reflections in relation to Question 3

My doctoral thesis deals with the origin of the large amounts of naturally formed organo-chlorine present in soil. To my surprise, I found a series of articles from the late 1950s and early 1960s on an enzyme called CPO that catalyzes formation of reactive chlorine. I figured that if this type of enzyme existed in soil it would produce organo-chlorine, and could thus be the source (or at least one source among potentially others hitherto unidentified sources). The enzyme belongs to a group of enzymes, which are called peroxidises (because they use hydrogen peroxide). In a textbook on soil enzymology, I found a recipe for extraction of peroxidises from soil. The remaining problem was how I would go about testing if soil contained CPOs or not. There was a standard assay for CPO but it was based on spectroscopy which wouldn't work on soil solutions, so instead I developed a method based on gas chromatography. To make a long story short, I combined theories from biological chemistry with techniques from soil enzymology, modified by integrating it with techniques from analytical organic chemistry and I analyzed the findings in light of recent findings in environmental chemistry. Biological chemistry is a field that focuses on finding naturally produced compounds that are useful, mainly focusing on pharmacologically interesting compounds and researchers in the field work in very clean laboratories. In contrast, soil enzymology works with soil, which is anything but clean and researchers in this field try to figure out which enzymes exist in soil, what they do and which organisms they come from. These disciplines ask different questions, use different techniques and they publish in different journals. To answer my question, I needed to combine theories and techniques from these traditions.

—

For my dissertation research, using knowledge from different disciplines revealed new insights into the challenge of how to adapt conservation policy. On the one hand, insights from the ecological and global change literature were crucial to understanding the scale of the problem and the very real impacts and feedbacks of various forces on both patterns and process of non-human elements of "nature". On the other hand, insights from anthropology and social-theories of science showed how prevailing conceptions of non-human nature have material impacts for the policies and management alternatives that are chosen. Together, the synthesis, integration and application of these insights combined for a greater sum of their parts. Without attention to the biogeophysical dimensions we would not appreciate the magnitude and dynamics of the biogeophysical forces acting on the non-human world. Without the socio-cultural insights and data, the research would have critically failed to address the social dimensions involved in how various policy alternatives related to managing "nature" are formulated and evaluated.

For my dissertation, the integration of knowledge from different disciplines applied to both concepts (as above) and also methods. As part of a mixed methods dissertation I conducted both in-depth semi-structured interviews, and also ethnographic participant observations. Had I not done the ethnographic data collection as a complement to the interview data I would not have adequately understood the tension (experienced by specialists) that a full realization of the impacts of climate change has for biodiversity conservation decision making. This provided important insights into understanding why conservation adaptation decisions had unfolded as they had.

Shannon Hagerman, September 2009

The purpose of my doctoral thesis was to compare and integrate different approaches to a practical problem; selecting areas for protection in the ocean. Therefore my thesis is inherently interdisciplinary. If I excluded one of the two main approaches I used, my work would no longer have been as novel or innovative. My thesis would have been very different – instead of the breadth I included in my work, I would have had to focus on depth within one of the approaches. Had I excluded the community-based approach, my work would have become primarily ecological; had I excluded the science-based approach, my thesis would have become primarily anthropological.

On a more personal level, I chose my topic because it allowed me to develop skills in different disciplines. The breadth I gained from using an interdisciplinary approach was part of the appeal for me when deciding to undertake my PhD. Not only did I gain breadth, I had to have sufficient depth of knowledge for each of my chapters to create an independent publishable article.

Natalie Ban, May 2008

It's difficult for me to think about Question 3 without thinking about both my aims. My academic aim was to combine different disciplines' models of behaviour into a single integrative model that could explain why people made decisions about energy. But this model was just a tool to help me reach my ultimate aim: designing policies to influence individuals' energy choices (towards more efficiency).

I figured that the more knowledge or models from different disciplines that I used and tested, the better my integrative model would be. And so if I excluded any given discipline from my research, I risked missing out on some important determinant of behaviour, some hidden source of influence. In terms of my academic aim, this might not have been so bad. I reviewed so many different disciplinary models that it became difficult to avoid them getting jumbled. It was an academic challenge enough to seek ways of integrating just two of them: say, behavioural economics and social psychology. But I didn't know how excluding a discipline might have affected my ultimate aim: what if that discipline held the key to an effective behaviour change policy?

Charlie Wilson, December 2008

My interdisciplinary research drew on and blended knowledge from several fields of study including: environmental policy; business strategy; and forestry management. I employed a mixed-methods comparative case study approach that included an in-depth review of primary and secondary sources; 120 semi-structured interviews; and the triangulation of the data with observational information gained by attending several industry and academic conferences. Using knowledge from the different disciplines to interpret and evaluate the data not only strengthened my research, it also made it possible. I drew on political science theory to understand the role of the state and the range of regulatory approaches; the business strategy literature to evaluate corporate motivation and market dynamics; and studies in forestry management to help contextualize and interpret the significance of the evidence and findings.

Jane Lister, 2009

Notes

1   Shannon Hagerman's PhD thesis can be downloaded at https://circle.ubc.ca/dspace/handle/2429/7903 or http://en.scientificcommons.org/50560592
2   Natalie Ban's PhD thesis can be downloaded at https://circle.ubc.ca/handle/2429/1275 or http://en.scientificcommons.org/46175251
3   Charlie Wilson's PhD thesis can be downloaded at https://circle.ubc.ca/handle/2429/2388 or http://en.scientificcommons.org/46174194
4   Jane Lister's PhD thesis can be downloaded at https://circle.ubc.ca/handle/2429/7212 or http://en.scientificcommons.org/50560753

5   See, for example, Liora Salter and Alison Hearn 1996; Andrew Barry *et al.* 2008; Julie Thompson Klein 1996.
6   Gunilla Öberg 2002.
7   The "price elasticity of demand" is a measure of how the demand for a particular good or service (say, heating a home) varies with price. Typically, if the price of heating a home rises, heating demand falls. The "long-run" means that this elasticity is measured over a long enough timescale for technological changes in response to the higher prices to be factored in, for example if the heating boiler had been replaced with a more efficient version. "Socio-technical systems" are the combination of technologies, infrastructures and social norms that provide the context for our everyday lives and decisions. Behind our decision to set the thermostat to 20°C is a whole system of physical and social structures, from boilers and gas pipelines, to energy utilities, homebuilders, energy prices, and an expectation of our household (including ourselves) that we can hang out indoors without our coats on. It is ultimately all these social and technological factors that "structure" or constrain our decision about heating. This in turn defines our responsiveness to changes in the price of energy, and so the "price elasticity of demand".

5 Why do you interact with society?

Many students who engage in interdisciplinary work wish to interact with people and organizations outside academia as part of their research. Some work in close collaboration with society because they want to provide support for decision-makers or influence decision-makers in a certain direction. Others wish to inform the public, and some believe that interaction leads to better solutions[1] and yet others because it is demanded by the outside society (such as funding agencies). Interaction between academia and society is as old as the universities. However, numerous interdisciplinary scholars do not desire to make interaction with the world outside academia part of their academic work, even though they conduct research across disciplinary borders. Such researchers work for and with their peers: they publish research articles or books which are read by their peers, they participate in academic conferences where they present their research for their peers and they write research proposals to funding agencies that use peer-assessment as the key tool for evaluation. This type of researcher interacts *only* with his or her peers and the peers are academics who conduct research in areas that are academically relevant for the work these researchers conduct. In contrast, transacademic scholars strive to interact with people outside academia in addition to their academic peers. In comparison with purely academic scholars, transacademic scholars must thus learn how to juggle even more diverse expectations.

Academic knowledge and decision-making

The interaction between academic research and decision-making in society has been studied for decades. A crucial lesson from the science-policy literature is that research, or any kind of information, plays different roles in

Interdisciplinary Environmental Studies: A Primer, 1st edition. © Gunilla Öberg.
Published 2011 by Blackwell Publishing Ltd.

different contexts. In non-complex areas with shared values, more information may function as simple and straightforward decision-making support.[2] The fuel gauge in a car can be used to illustrate this point: the position of the gauge is undeniably useful when the driver wants to decide whether or not it is time to purchase more fuel. Most situations are, however, more complex than the fuel situation described above. Complex issues such as nuclear power, climate change, conservation, land-use and pesticide use are impregnated with values, and different values form the basis for different political positions. As summarized by Roger Pielke Jr in his book *The Honest Broker*, factual information is not likely to change someone's position, in complex value laden situations. In such situations, it is crucial that the underlying values are clarified and made an explicit part of the discussion.

The belief that more information will always render better decisions is still held by a majority of issue experts, although studies in science policy have shown repeatedly that reality is far more complex. One reason for the unawareness of the context dependence of information is that only a fraction of undergraduate (and graduate) programmes pay attention to the relationship between knowledge provision and decision-making.

When you provide decision-making support, it is crucial that you are able to recognize the limits of your academic studies and to distinguish between situations where information is likely to compel straightforward action (as in the case with the fuel gauge) and situations where it is not. In other words, between situations lacking major value conflicts and situations with conflicting values. You are more likely to provide useful support if you scrutinize and reflect upon your role, your values and the values of the discipline(s)/research field you draw upon in situations that are complex and value-laden.

In complex situations with conflicting values, values are often obscured by purportedly objective, scientific arguments. Scientific arguments are "cherry picked" to support a specific political position. Science is "politicized". In such situations it is often more fruitful to use your abilities to identify *alternative* policy options than to act as an issue advocate. By identifying alternative policy options you may help create a number of scenarios, which has the potential to identify policy options that go beyond the value conflicts. Roger Pielke Jr calls this approach to act as an "honest broker":

> It is important to recognize that the Honest Broker of Policy Alternatives is very much an ideal type. In practice, having a truly comprehensive set of options would be overwhelming if not paralysing. And any restricted set of options will necessarily reflect some value judgments as to what is included and what is not. But there is a difference between providing a single option – issue advocacy – and providing a broader

set of options, particularly if the latter reflects a range of valued outcomes.

Pielke 2007, p. 142

Acting as an honest broker is generally more fruitful than acting as an issue advocate. It becomes highly problematic when researchers act as expert advisors but provide biased information without clarifying that this is the case. It is, however, *not* problematic to be an issue advocate if you act as an engaged citizen and *openly* state your value base. Being an engaged citizen can never be wrong, as it is a prerequisite for democracy and, like everyone else, researchers have values.

If you aspire to conduct studies in support of decision-making, it is crucial that you understand why it is impossible to solve value conflicts with more information. Not least because you are likely to encounter scholars who are unaware that it is impossible. When you act as a scholar providing decision-making support, it is advisable to learn about the value conflicts of the issue at hand, the values embedded in the information you provide and how this information is imbricated in the various value conflicts you have identified.[3]

Studies on science-policy interaction are published in books[4] and journals, such as *Social Studies of Science* (Sage Publications Ltd), *Science in Context* (Cambridge University Press) and *Studies in History and Philosophy of Science* (Elsevier Science Ltd). Unfortunately, most scholars who provide policy support or interact in other ways with society outside academia are not familiar with this literature, which is understandable as it is filled with jargon and quite impenetrable for the outsider. When read, the literature is often misunderstood, not least because it gives the impression that researchers deliberately act as biased issue advocates *pretending* to provide value-free advice. This is clearly a misunderstanding. Years of interaction with scholars who provide decision-making support,[5] in complex contexts such as acid rain and climate change, have led me to the conclusion that many scholars are unaware of the context dependence of scientific data. Hence, most scholars who act as stealth issue advocates[6] do so unintentionally. With a few notable exceptions, they honestly seem to believe that they provide value-free, objective information.[7] As discussed in Chapter 2, this seemingly naïve understanding is due to the fact that many scholars lack education with regards to knowledge-producing practices, questions that traditionally are handled in the humanities. If your background is not in the humanities, I advise you to learn basic "humanities jargon" so you at least can read (and understand) some of the key literature. In addition, you should find a broad minded person with a solid humanities background to collaborate with and to help you increase your meta-cognitive abilities.

Who participates in which part of the study and how? (Question 4)

Interaction with society outside academia is often delimited to research communication. The researcher focuses on how to communicate results to the broader public or, as in the case above, to decision-makers. There is a risk that research, which has no interaction with society other than reporting research results, will not grapple with what the real problems are and thus focus on problems of purely academic interest. Many scholars therefore choose to interact in two-way communication with various stakeholders and extra-academic experts to gain a deeper understanding of the issue. It is common to use consultation to achieve a deeper interaction, for example by interviewing people or sending out surveys to gain a better understanding of their viewpoints, perceptions and understandings. A consultative approach has the benefit of bringing a deeper understanding of societal needs to the table, provided of course that the consultation is carried out with rigour. It is surprisingly common that scientists with no training in stakeholder consultation methodology carry out poorly constructed surveys and interviews. It seems common to use different measuring sticks for rigour for the collection of information from stakeholders and extra-academic experts and other forms of data collection (Chapters 6 to 8). Collecting information from the public, stakeholders and experts is one of many ways to collect data. Non-rigorous consultative work bears many risks; among other things, it may contribute to even greater misconceptions of societal needs than does a pure research communication approach.

When using a consultative approach, the people you interact with are not directly involved in the study, and their role is to act as your consultants. There are scholars who argue that a consultative approach is insufficient in bridging the gap between academia and society, that only in exceptional cases will it lead to useful results, and that a participatory approach is necessary if you wish to produce socially robust knowledge.[8] A participatory approach means that extra-academic partners are involved in the whole research process, through definition of the research problems, design and implementation of the research, and interpretation, as well as use, of the results. This form of research is thoroughly discussed in the literature, for example under the terminology of "community based research", "action research" and "participatory research". One major challenge is that consultative as well as participatory studies are time-consuming and therefore demand more planning than less interactive approaches.

Whichever way you choose to interact with extra-academic society, it is crucial that you familiarize yourself with methodological and theoretical challenges outlined in the literature. There is no doubt that research which draws upon a deep interaction with societal partners is more likely to produce knowledge which is of direct use to the partners who are involved in the process. On the other hand, there is a risk that studies conducted in close interaction with societal partners result in stealth issue advocacy if the researchers are not sufficiently reflective and informed with regards to risks and benefits of the chosen approach. One way to decrease the risk of contributing to stealth issue advocacy is to clarify on what grounds you choose to interact with certain stakeholders or experts.

It appears evident that it is easier to understand the essence of a problem if you talk to those who have a stake in the issues at hand. But how do you determine who these people are? When choosing who to interact with, it is crucial to bear in mind that complex societal issues of high contemporary concern are usually highly politicized and the way a problem is defined will inevitably influence the outcome. And different people have different understandings of the problem. One group might define a problem in a certain way, which determines how the problem should be handled, which in turn might be the root of yet other problems (for other people). The ones you interact with in the problem formulation process will have a huge impact on the way you define your research problem, which in turn will determine what you will study and thus what issues you will address and which issue you will leave out.[9] It is especially important to reflect upon the choice of partners when you deal with value-laden issues such as climate change, land-use or transportation. In dealing with such issues, it becomes crucial to avoid the "just-do-it" attitude (Figure 6, p. 31), as such an attitude risks making you an unwitting stealth issue advocate.[10] It is crucial to find ways to welcome and encourage open, critical and balanced analyses and discussions on research results, interactions with extra-academic participants and the use of science in policy. In this context, the question of how the choice of participants colours the outcome of research efforts is crucial.

Embedded in Question 4 are sub-questions such as "Relevant for whom and for what reasons?" Hence, when you assess your work, a crucial question is "With whom do I interact?" or perhaps more importantly "Whom have I chosen to exclude and why?"

Commissioned research is a special case of interaction. Someone outside academia wants something done. There was perhaps a call for proposals and your research group was invited to compete for funding, or the commissioners might have contacted your research group as they were of the understanding that this particular group had the capacity to help them with their problem. In this kind of relationship, the commissioners are, most of the time, involved in the problem formulation process, but thereafter the information flow is most commonly a question of research communication: the research group meets with the commissioner and provides information about the progress of the work. Commissioners are seldom directly involved in interpretation and analysis. They have commissioned work from you and you are expected to deliver a product. Interaction is often limited to the start, the end and a few reporting incidents along the way. What happens if your findings are not to their liking? Will they pull the funding? What happens if you see something immensely interesting that your partners find uninteresting? Which track will you follow? The strength and the weakness of academia is our appetite for theoretical analyses, which often is perceived as unnecessary navel-gazing by outsiders. What is your strategy to keep a healthy critical eye and still be useful to your extra-academic partners? I would say that the solution most often is to clarify the expectation from both sides at the onset of the project. In addition, it is often helpful to separate the academic products (e.g. a doctoral thesis or an academic paper published in a peer-reviewed journal) from products aimed at your partners. Take some time to understand the different genres you need to master; hybrid documents that lie somewhere between an academic piece of work and a one-pager to the CEO of a company are not useful to anyone.

Reflections in relation to Question 4

The core of the empirical data for my doctoral research was semi-structured interviews with specialists in the field of climate change impacts and biodiversity conservation (including ecosystem managers outside academe). I engaged with interviewees in a consultative sense: they provided insight into their views on this particular topic. The nature of my interaction with the specialists was to learn their views (through interviews) and subsequently to provide an opportunity for them to give feedback as to my interpretation of their views (through preliminary internal review of a journal article in preparation for publication). While I sought feedback from interviewees, this work was not participatory. Participants did not aid in the formulation of research questions, they were not involved in the design and although they were given the opportunity to provide feedback on my interpretation of their views, my intention was to include feedback not in the editing of the paper, but rather as an addition empirical contribution in the sense of experts of this type (affiliation/world view. ...) tended to object to our interpretation of. ...

The question of who did not participate is crucial. In the case of my research the pool of interviewees was selected to represent key specialists in the field and to understand the gap between what scientists were stating off the record and the policy proposals that were advocated in the peer-reviewed literature. Stakeholders who would potentially be impacted by the eventual development/realization of various proposals were not interviewed at this phase of the research although this is the priority and logical extension for subsequent research.

Shannon Hagerman, September 2009

—

Part of my PhD research involved carrying out a community-based approach to prioritizing places for protection in the ocean, for which I developed collaborations with two indigenous groups in British Columbia, Canada. I focused on indigenous people because in Canada they are the only group that has constitutional rights to fish for food, social and ceremonial purposes. I also focused on indigenous communities because they are becoming increasingly proactive in planning their marine areas. Formalizing the collaborations involved discussing the research ideas with the local government – the indigenous chief and council – and getting their permission to carry out the work in their territories.

My methods of involving the indigenous groups approach that of a partnership as defined in this book. However, the indigenous groups were not involved in all phases of the work. In particular, I developed the research questions and approaches prior to collaborating with the indigenous groups. These were then adjusted based on discussions with key contacts. Therefore the approach taken was not a pure partnership.

My research with the indigenous groups consisted of three phases: (1) establishment of research collaborations; (2) semi-structured individual interviews with community members; and (3) feedback from the communities about marine conservation preferences obtained through the interviews. I used a snowball approach to interview people who are, or have been, active users of marine resources, and/or who have a particular depth of knowledge about the marine environment. I chose this focus because these people would be most affected if management changes were implemented. I stratified the snowball approach to ensure that we interviewed participants from all clans.

Natalie Ban, July 2008

—

My research question arose from the difficulties that governments, energy utilities, energy service providers and others were having in encouraging, influencing, or cajoling homeowners to invest in energy efficiency measures. So these "social actors" participated indirectly in the genesis of the research, and were very directly the ultimate beneficiaries (I hoped) of my research findings.

Their most important role, however, was in the data gathering. I could happily spend years reviewing the academic literature on behaviour, and behaviour change interventions, and coming up with new, shiny and even integrative, interdisciplinary models. But I wanted to test which models worked. For that I needed data: on household energy use, on efficiency measures undertaken (or not), on responses to policies, and on a whole host of psychological and contextual factors ranging from values, social norms, and attitudes, to indoor temperatures, the number of retired people in a home, and boiler types. Essentially, I needed to know about households who were the customers of energy service providers, the ratepayers of energy utilities, and the targets of government policies. So I worked with these "social actors" to come up with my sample design, and then to collect the data. For example, a big chunk of my data came from surveys administered to the customer base of an energy efficient home renovation company. I used to go down to their offices once a week to work on this sampling, and extensively used proprietary information belonging to this company (within all data privacy constraints, of course).

Charlie Wilson, December 2008

—

My research interest and questions were partially formulated by my experience working as a forest certification practitioner. I knew that the world was not black-and-white. Forest companies were taking on a public policy role and governments were adopting private forest rules. I wanted to understand the governance theory but also realized that my research needed to be transacademic – it was critical to go beyond the academic literature in order to understand how and why forest certification governance was actually occurring. Therefore, beyond conducting extensive data gathering interviews, I designed my research to incorporate contact with experts, practitioners, and scholars (outside my institution) at several critical stages throughout my PhD program. This "checking-in" not only helped to guide my research but also gave me confidence that I was on the right track as I charted the uncertain interdisciplinary waters. In particular, I made sure that I regularly attended forestry conferences (with industry and policy-makers) and also participated in several academic workshops with leading scholars from political science, business strategy and forest management disciplines. The interaction provided me with critical feedback on my research questions (year 1), research methodology (year 2), and results (years 3 and 4). For example, in presenting my research design and early results at an academic workshop at Dartmouth College, I was able to listen and consider the advice of the leading business school and political science scholars debating the merits and drawbacks of my study isolating "causation" vs identifying "influencing factors" of co-regulation. Participating in this debate advanced my understanding of the different methodological approaches and helped to focus and strengthen my research design, analysis, and results.

Jane Lister, November 2009

Why do you interact with society? (Question 5)

There may be many reasons why you wish to interact with people and organizations outside academia but if these interactions do not render a deeper understanding of the issue at hand, there are good reasons for you to think twice. Staying focused on your aim will help you decide who to engage with. You should consequently interact with various partners and stakeholders in a way that strengthens your study. One way to start may be to ask:

a) How does this interaction (Question 4) bring leverage to the study in relation to its aim? (Be specific, e.g. with whom, when and how)

This question may be difficult to answer and it is usually easier to start with the following question:

b) What would be lost if something was omitted? (Be specific)

Reflections in relation to Question 5

> The purpose of my study was to better understand the challenge of adapting conservation policy to the impacts of climate change. My approach was designed to examine the persectives of specialists in the field indcluding applied ecosystem managers actively involved in conservation decision-making. While my dissertation project was not strongly transacademic (its aims didn't require it to be so), plans for future work will absolutely include working with stakeholders so as to better quantify some of the implementation challenges involved in adaptive policies for conservation.
>
> *Shannon Hagerman, September 2009*

> The purpose of my study was to compare different approaches to prioritizing places for protection in the ocean. My community-based approach was designed to get perspectives from within the community, and to gauge the level of agreement amongst community members. Omitting any aspect of my approach would have resulted in a much weaker study that probably would not have achieved the aim. In particular, had I omitted my collaborations with indigenous people, the

community-based component of my work would not have come to frui-
tion. The transacademic aspect of my work was therefore crucial to the
overall aim.

Natalie Ban, July 2008

—

My ultimate research aim was to contribute to the efforts of governments, utilities, energy service providers and any other "social actor"
involved in trying to promote residential energy efficiency. Working
with these different actors through my research helped create a sort of
community of interest that would be receptive to my findings. That's
not to say that if they hadn't been involved at all, I couldn't have called
them up and gone and presented my findings anyway. But it certainly
made sure doors were open, and in some cases it also made sure that
the interests or concerns of those social actors were included in the
research design. Although this was a double-edged sword. In one
instance, I was pulled towards some research questions useful for profil-
ing a research partner company's customers than test behavioural
models (my research objective). In another instance, I found my research
was helping another partner company verify or calibrate their own
records. Neither instances were big deviations away from my narrower
agenda, but it did require some balancing.

The most substantive leverage, however, was in the data gathering,
but this was a peculiarity of my research design. I wanted to sample
households right in the middle of home renovations. Random sampling
would have given me very small and unrepresentative sample sizes. By
working with a home renovation company, I could use their customer
database to identify the households I was after. It worked really well,
but I could never have done this if I had not had a research relationship
with this company. This ended up being formally articulated in a col-
laborative research agreement between the university and the company,
a piece of bureaucracy I would not recommend unless there is a significant intellectual property rights issue at stake.

Charlie Wilson, December 2008

—

An important aim of my research was that it be relevant and useful to
policy-makers and practitioners. It would have been guesswork to
achieve this goal without having regular contact with the forest certification experts from outside the academy. I found I made critical
adjustments to my work after my various transacademic interactions.
For example, although the academic case study tradition is to focus
in-depth on just 1–3 cases, I was guided by the leading certification
practitioners at a US forestry conference to examine all twelve of the
certified US states in order to produce a much more useful analysis. As

well, after sharing my early research design with forest policy-makers, I was encouraged to diverge from the forest certification scholarship tradition and also include the national certification programs in my analysis – not just the Forest Stewardship Council program. I incorporated their advice with the result that my research ended up providing a comprehensive rather than a partial account of forest certification governance.

Jane Lister, November 2008

A word of warning: Don't be snobbish

Do you wish to produce knowledge that is used by someone in "the real world"? This question is rhetorical. Few want to produce useless knowledge. In Chapter 4, I spoke of interdisciplinary snobbism, and the detrimental effects of asking "Is this sufficiently interdisciplinary?" The question above hides a similar form of snobbism as it implies that work that is not trans-academic is useless. Unfortunately there is no foolproof way to produce useful knowledge. There are examples of academics who couldn't have cared less about the usefulness of their work, but whose work has been of immense value for an immeasurable number of people. The fact that someone is indifferent to the usefulness of their work does not mean that the work will not or cannot be useful.

Some people like to engage with people outside academia. Others do not. The literature suggests that this to a large extent is a question of temperament. You do not need to be derogatory about others' work to conduct work that shines. So, I suggest that you leave the question of whether or not your research will be *more* useful than someone else's out of the picture and focus on why *you* wish to interact with society, whom you wish to interact with, and how this brings leverage to your study. While you can increase the probability that your results will be used/useful, you cannot control *how* they will be used. Whether or not something is useful is a tricky question.

I wrote a thesis focusing on organochlorine compounds in soil because I was concerned about toxicological and eco-toxicological effects of such compounds. I found that such compounds are produced naturally. My results have been used by the industry to argue that there is no need to regulate such compounds since they are natural. Even though the argument is obscure (no one would argue that we do not need to regulate mercury because it is natural), there is little I can do about the way they choose to use my results, except to participate in the debate and hope that my arguments are heard. Just before I defended my doctoral thesis, I got a letter

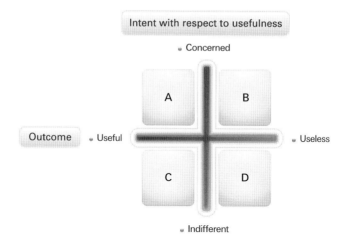

Figure 7 Illustrates that the outcome of a study is not necessarily dependent on the intent with respect to usefulness.

from Greenpeace, who were very upset about my results and accused me of being bought by the industry. I also accrued supportive letters from the pulp-and-paper industry. My thesis was clearly quite useful for the industry. It has also been useful for basic biogeochemistry in the sense that it has increased our understanding of biogeochemical cycles. In the long run, it might be useful for the regulatory process in the sense that it will bring about better ways to manage chlorinated organic compounds,[11] but this remains to be seen. My intention was to produce useful knowledge. So, I would place myself in the upper left-hand corner of Figure 7: I wanted to produce useful knowledge and the knowledge I produced has been used – but not in the way I hoped for. The lesson learned is that we cannot know and have little ability to influence who it is that might use the knowledge we produce, how they will use it, in what time frame and to what ends it will be used. As discussed in chapters 6–11, we can influence the rigor of our work, even though it remains to be seen if you or I will contribute to the greater good.

Notes

1   See, for example, Thomas Hughes' *Rescuing Prometheus.*
2   Roger Pielke (2007) uses the term "Science Arbiter" to describe the role assumed by a researcher who provides information that compels action.
3   More about objective, value-free information, in Chapter 2.
4   See, for example, writings by Sheila Jasanoff, Silvio O. Funtowicz, Jerome R. Ravetz, Bruno Latour and Helga Nowotny, to mention some prominent authors in the field.

5   My understanding is based on substantial experience of multi-participant projects aiming at providing decision-making support.

6   I have borrowed the term "stealth issue advocate", meaning someone who officially is an unbiased policy advisor but in reality acts as an advocate for a particular issue – in other words an issue advocate in disguise – from Roger Pielke (2007) *The Honest Broker.*

7   I discuss the notion of value-free, objective information in more detail in Chapter 12.

8   See, for example, Sherry R. Arnstein 1969; John Robinson 2008; Arnim Wiek 2007; Helga Nowotny *et al.* 2001.

9   More about the problem formulation process in Chapters 6, 7, 9 and 10.

10 See, for example, Roger Pielke Jr 2007; Liora Salter and Alison Hearn 1996; Julie Thompson Klein 1990.

11 The management of organic pollutants is actually impregnated by the obscure perception voiced by the industry that natural compounds are less problematic than anthropogenic ones, which complicates the management of such compounds.

6 Rigorous but not rigid

You may already have met someone who claims that interdisciplinary work will not and cannot be as deep as advanced disciplinary studies. Claims like that make it appear as if there is a clear-cut line between "deep" disciplinary and "shallow" interdisciplinary work. Rather than a description of reality, this is however a reflection of the tendency to use terminology with positive associations for the familiar (deep) and negative associations for the unfamiliar (shallow). No doubt, deep insights may be born within a traditional field as well as in the meeting between traditions. Already in 1969, Donald T. Campbell addressed the problem of breadth and depth of interdisciplinary work:

> Too often in discussions of interdisciplinary training one hears calls for breadth, for comprehensiveness. Too often we attempt the production of multidisciplinary scholars, professionals who have mastered two or more disciplines, rather than interdisciplinary specialists. This orientation I parody as the "Leonardesque aspiration": the goal of creating current-day Leonardos who are competent in all of science. As a training program, it is bound to fail in one or two directions. At its worst, it produces a shallowness, a lowest common denominator breadth, an absence of that profound specialization that is essential for scientific productivity. At best, it is evaded in the direction of the interdisciplinary narrowness here advocated.
>
> *Donald T. Campbell 2005, p 6*

The meaning of the word "depth" can be debated. My take-home lesson from Campbell is that a credible academic study must be rigorous, whether defined as disciplinary or interdisciplinary, where the aim of the latter is to fill unidentified gaps. Rather than being a distinctive mark of research within an established discipline as compared with research conducted across traditional scholarly boundaries, rigour may well be used as a tool to distinguish good research from bad, whether conducted in a traditional

Interdisciplinary Environmental Studies: A Primer, 1st edition. © Gunilla Öberg.
Published 2011 by Blackwell Publishing Ltd.

setting or not. Criteria used to evaluate the depth/rigour or scholarly quality of work should thus not differ from criteria used to evaluate the rigour of academic work carried out within a given discipline. The problem is, however, that "High Quality" is shaped differently in different disciplines. A great number of differences are due to different disciplines using language and illustrations (figures, pictures, tables, etc.) in utterly differing ways. Different terms are sometimes used to describe the same phenomenon and the same terms are sometimes used to describe different phenomena. Often differing formats are used, such as headings, where to write what, and how to take care of references. Some of these differences are incommensurably rooted in different epistemologies, while other differences are more a question of culture. Hence, when involved in activities that span traditional scholarly borders, you must acquire consciousness of and respect for variations among research procedures. A widened understanding of colleagues' research practices is often the key to an increased ability to distinguish bad research from disciplinary differences.

The key is to identify criteria for assessment of *common* attributes of credible academic work: criteria that are independent of academic traditions and go beyond disciplines. In Chapters 7 to 11, I discuss five criteria (one per chapter) that are common attributes of credible academic work but are shaped differently in different disciplinary contexts. The crux is to learn to see beyond the shape.[1]

Most disciplinary-based scholars have a rather clear understanding of the demands of a credible study, even if these demands seldom are clearly verbalized. The difference between a homogenous and a heterogeneous academic environment is that it is possible to rely on tradition in the former; quality comes with the territory.[2] In such an environment, a student acquires the competence to conduct a good study mainly through apprenticeship: by spending time in an environment where teachers, supervisors and other researchers use the same type of methods, models, concepts, theories and thought styles that the student learns to manage and later use. Texts that are produced by teachers are of a similar format to the ones the student reads and eventually will write. By being entrenched in tradition, so to speak, one learns how to demarcate and anchor a study in a credible manner. The relevant literature is more or less given as well as the perspective used to assess collected empirical data critically. The agreed-upon norms go with the territory and the learning process is mainly driven by spending time in the environment rather than by conscious reflection.

In an interdisciplinary environment, various traditions meet and it becomes harder to acquire an ability to distinguish bad from good without conscious reflection, not least of all since the final product is not a given. One of the major opportunities in conducting studies outside traditional borders is that the audience is less clearly defined. The work has the potential for being read by scholars of varying backgrounds, which, however, means that the readers quite likely have different understandings of a "credible study". A study that moves across traditional academic borders therefore does demand more elaborated and exhaustive explanations and clarifications than a more traditional study does.

The fact that the environment often stimulates (forces?) students (and scholars) to revisit their tacit understanding of high quality is probably one of the reasons why so much exciting and creative research is done at the frontiers where traditions meet and mix.[3]

On quality assessment

Critical assessments are defining elements of the academic world: research proposals are examined by review panels, working papers are discussed and evaluated at seminars and conferences, manuscripts are examined in the peer-review process,[4] and research presented in books is criticized in seminars, in journals or in yet other books. In other words, research projects are, to various degrees, evaluated and assessed in all their stages from the initial proposal to the final stage when presented in articles, reports or books. One of the goals of an academic education is to help students acquire skills such that they are able to conduct credible studies and to distinguish reliable from less reliable information, or to put it more bluntly, to distinguish good work from bad. To achieve these goals, students are assessed by teachers, tutors and supervisors.

The assessment process is driven by explicitly and implicitly agreed-upon norms and procedures. Many norms and procedural issues are taken for granted and not considered relevant to reflect upon, and many of these norms and procedures are rooted in tradition rather than in conscious teaching and research practice. As long as you remain within a certain discipline, the implicit rules render few problems, since all those within the tradition embrace and follow the rules – even those who do not understand the meaning of the rules. However, if you move among disciplines it becomes apparent that the explicit as well as implicit norms and conventions vary quite a bit among traditions: credible studies come in many forms.

There are, however, common traits. Empirically-based academic traditions generally demand that a study is conducted in line with some sort of standardized gross structure. You are expected to have pondered questions such as: "What kind of information is going to be collected and how?", "How is the information going to be analysed?", "What theoretical framework is going to be used?", "What components comprise the research procedure?" and "Where in the final text shall the components be accounted for and to what extent?" Approaches to these questions vary strongly among disciplines. Some scholars expect a detailed description of the procedures, whereas others prefer more overarching, sweeping descriptions. Yet others demand detailed accounts for some of the basic procedural components at the undergraduate level but this demand is omitted during graduate studies or at the post-doctoral level as a scholar is expected to be familiar with the craftsmanship when the doctoral thesis has been successfully defended. The procedural rigour of a study is thereafter often hinted at rather than made explicit. Such hints give clear signals to the informed but are impossible to understand or even detect for the outsider. In other cases, some of the basic procedural components are boiled down to a cipher-like text and the rigour is accounted for by giving reference to supposedly credible sources. It is thus not surprising that students (and scholars) experience difficulties in acknowledging the rigour in other traditions' procedures. As a direct consequence, *it is easy to judge other traditions as less credible than your own*. Many have testified that an increased understanding of how rigour is achieved in traditions other than your own is a major key to successful collaboration across traditional borders.[5]

Confusing form and credibility – an example

As previously discussed in Chapter 2, the gap between the humanities and natural sciences is often used as the most obvious example of differences among research traditions. Undeniably, this gap is wide and difficult to bridge, but it can be bridged and, more importantly, there are other gaps that urgently need to be bridged but are hidden by the intense focus on The Gap. In this book, I point out traditional differences across The Gap, and how these traditions hinder the identification of common ground and thus collaboration. In the example below, I use the striking difference in the use of language in a natural science text and a text produced within the social sciences to illustrate how a pronounced difference in form may be confused with credibility (from both sides).

The language of the natural science text in the first quote in example 7 is aseptic and compact with a close to poetic shape in its strict format. In contrast, the humanities text in the second quote is exuberant, expressive and verbose – and much, much longer. The natural science text is larded with technical terms and the layman is probably already lost in the terminology in the first sentence. The social science/humanities text uses fewer technical terms but the layman is also lost in this text, perhaps not as fast and probably not primarily because of the terminology but because of the reasoning, which is difficult or even impossible to follow for the uninitiated although the vocabulary is familiar. It therefore takes a longer time to realize that you have been excluded from the humanities text, whereas the natural science text excludes the layman from the very start.

Example 7
Different writing styles

Conclusion: Our results show that fungal CPOs are potential chlorinators of lignin in plant debris and soil and may thus account for some of the high-molecular-weight organochlorine residues that occur naturally. Because they are ligninolytic, CPOs may also contribute to the slow breakdown of lignin in soils. CPO activity (1) and H_2O_2 (11, 16) both occur in soils, and Cl^- is of course ubiquitous in this environment. CPO catalyzed ligninolysis provides an interesting biological parallel to the well-known industrial use of chlorine for wood pulp delignification (4).

*Patricia Ortiz-Bermudez **et al.** 2003, p. 5017*

Conclusions: Alternative conceptions of science and communities
Much like the geopolitical constructs of "North" and "South", the epistemic community concept is useful. However, like the former, it must be used with care, with sustained attention to the important divisions they fail to capture and threaten to obscure. Rather than a single, cohesive, transnational "community" united by a shared professional ethos and by scientific and policy-related concerns, this study reveals a complex domain characterized by transnational networks and cognitive convergence, but also important differences. Persistent lines of division exist – sometimes under surface appearances of shared frameworks of understanding and policy action – testimony to the precarious and unstable nature of any apparent consensus on global environmental problems and the associated knowledge (Brosius 1999; Escobar 1999). Divisions continually replicate themselves as one looks from the transnational to the national and even the individual level. On second glance, not only does the transnational epistemic community appear internally fractured along geopolitical lines; important fractures also reveal themselves at the national level and even within the subjectivities of individual scientists, at the most intimate level of personal commitments and understandings of self and the world. Even so, patterns appear in

Continued

the fractures, conditioned by history and socio-economic realities. The scientists identify as Brazilians, as citizens "of the South" and of a formerly colonized country. They are transnational locals. They are more local than transnational, perhaps, in that their discourses consistently evoke the "North-South divide", even as they know themselves to have been intimately shaped by their foreign (overwhelmingly US) educations, work experiences, and associated personal relationships. The recurrence of the North-South construct in Brazilian scientists' discourses bears testimony to the reality glossed over by epistemic community theorists, the extent to which climate science and geopolitical conditions co-produce each other. Privy to this insight and less advantaged by the associated politics, Brazilian scientists and policy-makers see the shadow side of universalizing discourses about science. Matters such as race, ethnicity, and level of development shape their understandings and experiences of international science, and they perceive prevailing international climate related knowledge as typically reflecting developed ("Northern") countries' assumptions, agendas, and interests. Such perceptions, and the differences they reveal, add complexity to the dominant, purified images of science, representations that overlook important questions about the intersection of science, culture, power, and politics.

Myanna Lahsen 2004, pp. 170

Excluding the layman is to many a sign of credibility. When I was to defend my doctoral thesis, one of the professors at the department burst out: "Gunilla, the way you write, one might think this is simple." Although clearly not meant as a compliment, I took it as such. Sadly though, the comment reveals what many hold true: good research must appear complicated. A research text must undeniably contain specialized terms and advanced reasoning, but the closing of the text to the uninitiated is, in both of the above examples, to a certain extent a question of mannerisms. If we are to bridge gaps among disciplines, we must eradicate jargon and other unnecessary exclusion mechanisms but still keep the depth and rigour: a complicated balancing act worth elaborating in collaboration with colleagues – not least because we are blind to our own jargon.

A commonly held view among my natural science colleagues is that a *long* text reveals that the author has not done the homework; in other words, the text is not credible – it is not ready, it needs more thinking. In contrast, my humanist friends often argue that a *short* text reveals that the author has not done the homework; in other words, the text is not credible – it is not ready, it needs more thinking. I claim that it is impossible to use the length of a text to judge whether or not it presents credible research. Different traditions have different advantages as well as drawbacks, and they are more or less suitable in different contexts. The length is an outer feature and cannot be used as a criterion of credible research. We must learn to look beneath such outer features to find common ground on a more fundamental level.

Sound research practice is a necessary requirement for credible studies (although it does not automatically lead to credible research), and a great deal of the common ground among disciplines can be found in the practice. A closer scrutiny and comparison of the various tradition-based criteria of *sound research practice* reveals that many of the embedded criteria are differing expressions of the same idea – the shell varies but the kernel is the same.

Communication

The success of any research project demands that the collaborators engage in fruitful dialogue.[6] In most cases, communication within a traditional academic environment comes more easily than communication between scholars of varying background. A student who wishes to combine or integrate knowledge from more than one discipline will consequently face communication challenges that students in a more traditional academic setting will not face.

By initiating dialogues among fellow student with mixed academic backgrounds, you are likely to achieve an increased awareness of how design and context shape and influence the results and usefulness of a study. This is an efficient way to sharpen your study. Various communication barriers hinder fruitful dialogue and create unnecessary frustration. An increased awareness of key concepts that often cause misunderstandings is one way to decrease frustration and enable collaboration. There are a number of concepts that are crucial to achieve academic rigour, such as "theory", "method", "evidence" and "analysis". Unfortunately, these concepts are particularly difficult to handle in interdisciplinary contexts and I will discuss each of them in later chapters. A worthwhile exercise to kick-start your interdisciplinary understanding of these key concepts is to identify three papers that in your view represent "interesting high-quality research".[7] Use four markers of different colour and mark text that describes theory, method, empirical evidence/data and analysis. Meet with a group of students who have done the same exercise and discuss your findings. I will return to these three papers in the following chapters, so keep them for further exercises.

I wish to draw your attention to four other terms that cause unnecessary confusion in interdisciplinary settings: "relevant", "interesting", "credible" and "useful". There are seldom reasons to ponder the differences among these concepts in a mono-disciplinary environment, since everyone to a large extent has shared academic interests where "credible" is used interchangeably with both "interesting" and "relevant". Anything of interest is also understood as relevant and the study will also be considered credible, provided the study is conducted in line with agreed norms.

In contrast, when you start conducting studies that in one way or another cross traditional academic borders, the differences between the three terms "credible", "interesting" and "relevant" become rather problematic. There is a great risk that misunderstandings will cause problems if you do not reflect upon these differences. My hope is that an increased awareness of the usage of these terms will help you in your problem formulation process.

Interest and relevance

One characteristic of sound research is that it is relevant in a wider perspective than the specific study in itself. The question is what perspective? In a group with shared disciplinary interests, "relevant" most often becomes synonymous with "interesting". Students and scholars in a traditional research environment usually have shared interests, and "relevant" is thus placed on equal footing with "interesting". A credible research project is thus often understood as "a study dealing with interesting issues from the point of view of our disciplinary community." As long as one stays within the same disciplinary community, the fact that interest is conflated with relevance does not generally cause problems. In the figure below, studies in the top left corner (interesting and relevant) are self-evident and easily defined in a disciplinary context, since things that are considered interesting by definition also are relevant and thus easily distinguished from issues that are irrelevant, as these, by definition, also are uninteresting (square D). The squares B (uninteresting but relevant) and C (interesting but irrelevant) are more or less empty, since things that are of no interest per definition are irrelevant.

In a disciplinary context, relevance is implicitly understood as anything that brings further insights to the given field. In an interdisciplinary setting, there is no given field, since any "given field" at a closer look takes the shape of a diffuse landscape. It is easy to intuitively take for granted that the things I find relevant (= interesting) are also understood as relevant (and interesting) by the others. A first step to avoid conflating "relevant" and "interesting" is to become aware that the very same research question can result in completely different studies. For example, the question: "What are the effects of forest fertilization?" will most likely spark rather different ideas in different disciplinary contexts. A forestry department is likely to design a study in which the incremental growth of the economically-interesting trees were to be studied. A department of soil microbiology might instead set out to investigate the influence on the degradation processes or the microbial community, whereas hydro-chemists probably would agree to study the effect of water chemistry in the run-off, and socio-economists might investigate the influence on the forest owners' economy in the short and long term. The macro-economist might instead try to elucidate the influence on GDP, whereas the department for ecotourism might initiate a study on the influence on outdoor life.

For students who are involved in an interdisciplinary project, it is in every respect important to bear in mind that there are always a number of possible ways to start from any given point. Conducting interdisciplinary studies is to enter into emerging fields, which to a larger extent than in a disciplinary study is a question of entering the unknown. More energy must therefore be spent to clarify, first to oneself (the supervisor and committee) and later to the reader, what the field is and why the particular study is relevant. In other words, whether or not a study is relevant depends on the scope of the specific study and the context the study is conducted in. In an interdisciplinary context, "interesting" is thus seldom self-evident. Issues that one person finds interesting others may find utterly boring, and issues that those others find exciting may appear incomprehensible to others. Still, if sufficient time is spent to understand why boring things make others excited, a mutual understanding may arise, not to the extent that everyone will be excited about the same thing, but to the extent of a shared understanding why various and differing issues are seen as relevant (and interesting) – and not least, that things that I find boring might be very relevant.

Separating "interesting" from "relevant" can actually be used to render the research process more effective, by revealing issues that the individual student finds interesting but are revealed as of no relevance when mirrored in the aim of the study in question (square C, figure 8). To separate "interesting" from "relevant" can thus be used to focus a study. All who have conducted a study know focusing is crucial. If you work in a larger project with several participants, the key is to find a common definition of the aim or the scope of the study and to critically scrutinize the relevance of the study or various sub-studies in the light of the formulated overarching aim. Engaging in dialogue with others to explain the relevance of whatever you are doing is most often a very healthy process, whether you are involved in a larger project with a number of collaborators or paddling along in your own canoe.

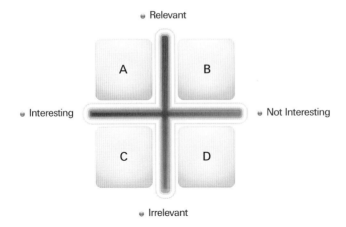

Figure 8 Illustrating that the relevance of a study is not necessarily related to whether or not it is interesting.

Credibility and relevance

The question: "What is credible research?" is constantly discussed and negotiated in the research communities. The debate, in seminars, workshops, conferences and research journals on what to consider as credible research is in itself held as a tool that enables the production of credible research – everything can and should be questioned, discussed and debated. The review-process has been criticized as flawed and coloured by subjectivity,[8] which has shaken those who believe that context independence is the core of credible research and the review process *guarantees* that credible, objective research will be produced. Other more pragmatic researchers (I dare claim the majority) conclude the open review process is the best we have, since it is far better than *ad hoc* criticism and whatever other alternatives are at hand. The latter group admits more or less openly that it is clear the process is not objective in the sense of being context-independent, neutral and value free. All types of research are, and must be, conducted from a given perspective, as it is impossible to take everything into account.

It does not come as a surprise to those involved, except perhaps for the newcomer, that peer-review is a rather subjective process. A paper judged as excellent by one reviewer may be doomed as slipshod by another, as exemplified by the citation below in which the first reviewer states that "the work appears to be sound and the conclusions reasonable," whereas the other reviewer expresses "serious concerns about the interpretations the authors are making." The third reviewer also suggest the paper be rejected, but not because of lack of quality, but because the reviewer believed the manuscript was not suited for publication in the specific journal:

Example 8
Assessment of quality and relevance

Additional Comments: this manuscript describes ... The work appears to be sound and the conclusions are reasonable.

(Anonymous referee no 1, 2004)

Overall comments: This paper is well written and the design and execution of this study was careful and thorough. I believe the data are also of high quality. However, I have serious concerns about the interpretations the authors are making. /.../ Because of these things, I think this paper does not contain sufficiently new or important scientific material or ideas to contribute significantly to existing knowledge.

(Anonymous referee no 2, 2004)

—

I thought the manuscript was well written and the authors present an interesting investigation /.../. It seems as if *ES&T* is not the best journal for this type of paper. Perhaps *Ground Water Monitoring and Remediation* may be more appropriate outlet.

(Anonymous referee no 3, 2004)

Hence, a paper or a book can be rejected either because of lack of quality (quadrants B and D) or because it is out of scope ((quadrants C and D); Figure 9). In this respect, peer-review resembles other instances of assessment in academia: assessment of student work, of research proposals and of applications for academic positions. A teacher evaluates if a student paper is of sufficient quality *and* if the work is within the scope of the course curriculum. For example, someone in charge of a course in ancient history will probably not accept a paper entitled "The role of soccer in World War II." The title bears no reference to the core of the course, which indicates that the paper most likely is outside the scope. It doesn't really matter whether or not it is a paper of high quality: if it does not fall within the scope of the course, the student will fail. The same goes for a research proposal or an application for an academic position. It is not sufficient simply to submit a superb proposal or to have produced large amounts of high-quality work: the proposal and the work must also be within the scope defined by the funding agency, the advertised position or the course curriculum. Rejecting a paper/proposal/application because it is outside that scope, in other words irrelevant from the given perspective, should

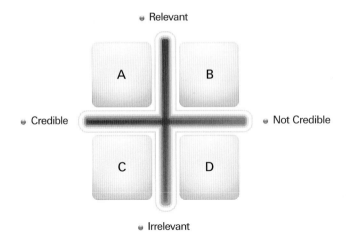

Figure 9 Illustrating that the relevance of a study is not necessarily related to its credibility.

consequently not be confused with the quality of the work – a paper/proposal/application can be of really high quality, but out of the scope of a given perspective. The fact that a study may be outside the scope of one perspective but well within the scope of another perspective should thus not cause a problem when the issue at hand is to judge whether or not conducted research is of high quality.

In a disciplinary context, people have a fairly similar understanding of the scope of a course curriculum, call for proposals or an advertisement for an academic position. The interpretation of the scope is guided by the overarching perspective of the discipline. The overarching perspective is, however, seldom explicit, but rather embedded in the research traditions and understood as a sign of credible research instead as a choice of perspective. The overarching perspective becomes the guiding principle when assessing quality – papers are rejected because they are outside the implicit scope. The concept of quality is tacitly guided by the paper's relevance in relation to the (implicit) overarching perspective.

When you plan and conduct interdisciplinary studies, it is advisable to discuss and clarify in what way the study is relevant. From what perspective? To what end? If the overarching aim is given by a course curriculum or call for proposals for example, make sure to carefully explain how your work is relevant in relation to the curriculum. It cannot be taken for granted that your interpretation of the curriculum is in line with the assessor's interpretation, and there is a great risk that you will fail unless you have provided a clear and concise as well as carefully elaborated (not necessarily the same as a lengthy explanation).

Notes

1   See Veronica Boix Mansilla and Howard Gardner 2003.
2   Drawing on Campbell's expression (2005), this is in part a result of the "ethnocentrism of disciplines".
3   See, for example, Nancy Sung 2003.
4   Peer-review means literally that the researchers are reviewed by their peers, i.e. other researchers.
5   See, for example, Veronica Boix Mansilla and Howard Gardner 2003.
6   David Bohm 1996.
7   I owe credit for this simple and elegant exercise to my friend and colleague Theresa (Terre) Satterfield: http://www.ires.ubc.ca/?p2=/modules/ires/profile.jsp&tid=25
8   See, for example, David Shatz 2004.

7 Marking your playground

Research projects have a tendency to sprout in all kinds of directions, seemingly by their own force. To avoid becoming overwhelmed, you need to mark your playground by making a number of choices: what will you study, with what method and in light of what literature?[1] Question 6 in the framework is designed to help you with this process: Is your study sufficiently and coherently demarcated? The process through which you set the borders of your study is actually initiated before you start to consciously design the study.[2] Many of these choices are unconscious, guided by tacit norms – which vary among disciplines. Increasing your understanding of how different disciplines handle choices will help you to transform some of your unconscious choices to conscious ones, which undeniably will help you produce a stronger study.

All disciplines demand that central choices are accounted for in the final text but they are accounted for in different ways. The fact that choices are accounted for in an unfamiliar way often causes researchers to deem texts from other research traditions "bad research." One reason scholars have difficulties accepting others' ways of doing things is that they are unaware of their own ways. Even though all disciplines emphasize the need to clarify central choices, few curricula stimulate awareness of how choices are made in the discipline and how this differs from other disciplines. Becoming aware of the traditions of your own discipline and becoming familiar with the traditions of others is an efficient way to handle some of the obstacles of interdisciplinary research. A fruitful way to transform this challenge to your advantage is to form discussion groups, select typical research texts from the disciplines represented in the group, and analyse, describe and discuss the choices.

In support of such dialogues, I point to four areas in which most disciplines agree that choices should be accounted for, even though the

Interdisciplinary Environmental Studies: A Primer, 1st edition. © Gunilla Öberg.
Published 2011 by Blackwell Publishing Ltd.

terminology as well as the extent and where in the text the explanations are expected to be placed varies strongly:

- framing (choice of perspective);
- choice of aim;
- choice of object/subject of study (how to operationalize the aim);
- choice of technique for collection of data.

It is worthwhile spending some time reflecting on these four domains of choice and how elaborate you would like the explanations of each to be and why; as well, discussing where the explanations should be placed will clearly improve any research text, whether written in an interdisciplinary context or not.

Framing

A new project can be born in many ways. It may be a conscious process where you start by choosing a perspective and then methodologically chiselling out the aim. It may just as well start with an idea that seems to come from nowhere, or it may be the case that you continue a thread with a long history. Irrespective of how your study is initiated, your perspective is consciously or unconsciously chosen early in the process. The process starts well ahead of the planning and continues throughout the project.

Choosing a perspective means that you choose the lens or the grid through which you will look at the issue, system, process or phenomenon. As discussed in the previous chapters, the choice of perspective is most often a tacit process that is deeply interwoven with issues of interest and relevance. Since interests vary among researchers as does the apprehension of relevance, different persons will approach the same problem from different angles. In other words, they may choose to study the same issue from different perspectives. For example, the same geographical area, say the Amazon rainforest, can be described from a socio-economic perspective, from an ecological-biodiversity perspective, from a transport logistic perspective, from a carbon budget perspective, from a gender perspective or from an equity perspective.

In some traditions, the perspective is implicitly guided by the overarching perspective of the discipline, whereas the process of choosing a perspective is more open and explicit in other disciplines. The perspective influences the whole research procedure, from the definition of research questions, choice of empirical material, and development of theoretical framework to analysis and interpretation to the form of the text.

In many disciplines, texts start with a background description, thus providing a rationale for the study. Some disciplines demand that choices made are clearly spelled out and justified in the introduction. Others expect a general framework in the introduction and demand a specific paragraph further on where the chosen perspective and other major choices are justified. Yet others do not explicitly spell out the choices. Instead the choices are implicitly shown through the line of reasoning rather than being clearly spelled out. Which format have you chosen? How will you justify the choices you have made? To help you to choose a format that suit your ends, I provide one example where the author has chosen to use an explicit approach when clarifying the choice of the perspective and one example of a study where the author has chosen to use a more implicit approach.

Example 9
Explicit choice of perspective and explicit aim

The starting point for this study is the importance of international co-operation on transboundary environmental issues. Within the co-operation there are different approaches to the problem, such as legal, scientific-technological and political-programme strategies, which complement one another.

The aim of this thesis is to study how the problem of pollution from diffuse landbased sources, especially agriculture, is dealt with regionally in the Baltic Sea area through the work of the organisations Helsinki Commission (HELCOM) and Baltic 21. More specifically, the roles of and the relationship between HELCOM and Baltic 21 will be studied. Furthermore, the possibilities and the difficulties in the practical co-operation as it appears from interviews will be dealt with. The study offers a Swedish perspective and concentrates on the views and experiences of Swedish representatives.

/.../ Both HELCOM and Baltic 21 imply co-operation on a state level, between governments, and are in that sense similar. This could facilitate the comparison of them, but it also means that the "outside" perspective of a Non-Governmental Organisation (NGO) has been left out.

Stina Karlsson 2003, p. 9

The example is from Stina Karlsson's masters' thesis on international co-operation and the author has chosen to focus on co-operation in regional environmental issues. It is delimited to a specific environmental problem (pollution from diffuse land-based sources), in a specific geographical setting (the Baltic Sea), in two specific political settings (HELCOM and Baltic 21), from the perspective of one of the involved countries (Sweden). Hence, five choices that frame the study are clearly spelled out.

Choices made in Example 9:

- *Overarching research field:* Co-operation in regional environmental issues;
- *Environmental problem:* pollution from land-based sources;
- *Geographical setting:* the Baltic Sea;
- *Political setting:* HELCOM and Baltic 21;
- *National perspective:* Sweden.

Example 10 illustrates a more implicit way of writing.

Example 10
Implicit choice of perspective and explicit aim

Introduction

Nitrogen is an important nutrient and it has long been the limiting nutrient in Swedish forests. /.../ Clear-cutting is a common method of harvesting in Swedish forestry. The negative environmental effects have been documented in several studies both in Sweden and abroad (Rosén *et al.* 1996; Ensign 2001; Nisbet 2001; Tonderski *et al.* 2002). In a clear-cut forest the soil temperature increases and the mineralisation and nitrification increases. In the same time the evaporation decreases due to the lack of trees and there will be increasing flow peaks. The disturbance in the forest leads to an increased amount of nitrogen leakage (Ring 1995; Tonderski *et al.* 2002). /.../ The dominating form of nitrogen in the leachate in a non-cut forest is organic, but in the clear-cut forest nitrate will dominate the leachate (Ring 1995; Bergquist 1999; Tonderski *et al.* 2002). /.../ Although these and other measures are taken in the forestry, nitrogen leakage is still a problem and it must be reduced. A way to decrease the nutrient transport to aquatic environments is riparian zones (Bergquist 1999). /.../ According to Bergquist (1999) the nutrient leakage from forest harvesting has not been considered as a big problem on water quality. /.../ Another problem is that there are several studies that only look at the period of growth and not the whole year (Bergquist 1999). If only the period of growth is considered, the nitrate reduction results can be misleading and the riparian zone left after clear felling underestimated.

The purpose with this thesis is to see if there is a significant difference in nitrate leakage between a clear-cut site and a reference site over the years, in the province of Halland, Sweden. Furthermore, to investigate if there is a seasonal variation in the nitrate leakage from the clear-cut area.

Peter Osgyani 2003, pp. 4–5

Peter Osgyani, the author of Example 10, chose to focus his masters' thesis on problems encountered when monitoring the environmental quality in the field. The aim of the study is justified by a number of interwoven arguments from an implicit perspective. The aim of the study is to investigate whether or not it is possible to monitor if border-zones in forests prevent nitrogen leakage in spite of a potential seasonal variation. Peter starts by justifying why it is interesting and important to study nitrogen in forest soils. As understood but not clearly spelled out in the text, the study is rooted in a specific desired outcome: incremental growth. The tacit line of reasoning is thus that forestry is one of the major industries in Sweden; that nitrogen is one of the factors that influence incremental growth, and that factors influencing incremental growth are of great interest as growth is the income basis for the industry. This is well known to all in the field and thus considered needless to spell out. Not one of the choices made is explicitly stated but many are clearly justified through the reasoning in the text. For example, Peter explains that it is well known that clear-cutting does influence nitrogen leaching (so he chooses to study the effect of clear-cutting on nitrogen leaching), that border-zones are used to decrease such leaking (so he chooses to study the effect of border-zones), and that there have been indications that the leakage varies seasonally. Peter explains that previous studies have been carried out in such a manner that it has not been possible to distinguish between seasonal variation and the effect of border-zones, which is necessary in order to evaluate whether or not border zones are efficient tools to reduce nitrogen leaking (so he chooses to focus on whether it is possible to distinguish these from each other). In this case, like the one above, a number of choices are spelled out, although in this case they are implicit rather than explicit:

- *overarching research field:* methodological problems while monitoring the environmental quality in the field;
- *environmental problem:* leaking of nutrients due to clear-cutting;
- *nutrient:* nitrogen;
- *management practice:* border-zones;
- *geographic setting:* forest in Halland (a region in Sweden);
- *underlying perspective:* incremental growth.

The style illustrated by Example 10, justifying through reasoning, is a bit seductive. For the uninitiated, the context and framing makes it appear as if the proposed study is the only logical outcome. The choices made are invisible. Persons who are used to reading this sort of text are however generally used to deciphering the text; in order for such a text to be credible, the choices have to be justified by well founded arguments embedded in the text.

Implicit or explicit?

As shown in the examples above, some disciplines expect that the choice of perspective is made explicit (as in Example 9), whereas in other disciplines, the choices are implicitly understood through the framing, as in Example 10. Researchers from "explicit-perspective traditions" often judge texts from "implicit-perspective traditions" as sloppy and non-credible since they do not contain explicitly stated proclamations of the choices made. On the other hand, researchers from "implicit-perspective traditions" often judge texts from "explicit-perspective traditions" as sloppy and non-credible, since the framing is structured in an unfamiliar way. The "explicit-perspective traditions" emphasize that choices are justified and clearly spelled out. This makes the texts longer and "implicit-perspective traditions" interpret them as unstructured and babbling.

In my experience, it is possible to avoid rather confusing and discouraging conflicts with supervisors, committee members and fellow students by initiating dialogues on the extent to which it is necessary to spell out, for example, choice of perspective in accordance with some traditions, or to let the choice be made clear through the framing, in accordance with other traditions.

Aim

Perhaps needless to say, but it is generally easier to conduct a study if you know what you wish to achieve. Surprisingly many set out to conduct studies without knowing why. Do you wish to illuminate, question, scrutinize, analyse or perhaps describe something? Formulating the aim is a process and, in most cases, the problem formulation process is ongoing during the entire project. You will find yourself rephrasing the aim and central research questions numerous times as new information is gathered via literature and own observations. It is a crucial and strenuous part of the research process and it is important to realize that the problem formulation process cannot be skipped. The more specific and clear-cut you manage to formulate the aim at an early stage (to be revised over and over again at later stages), the easier it will be to design the study. Many are those who pay testimony to the virtue of dedicating time and effort to the problem formulation process, exemplified by a citation from a graduate student at UBC:

> In the end, lessons learnt from this experience should help us in our own research areas by recognizing that objectives have to be clearly defined before undertaking any fieldwork, otherwise confounding effects in the data would delay getting (if) any useful results.
>
> *Roseti Imo 2007, with the author's permission*

Roseti participated in a course in which he had written a proposal jointly with some other students. When at the end of the course, reflecting upon the role of the problem formulation process, he expressed a number of lessons learned, among others to never again withdraw from the process. It was obvious to him that time won by withdrawing from initial discussions will be lost many times over at the latter stages of a project. It is difficult to be efficient if you do not know why things are done in a particular way. Interdisciplinary settings usually provide eminent support for the problem formulation process, since the differing backgrounds of the people will force you (i.e. help you) to formulate your aim in simple, clear-cut terms. The importance of formulating a clear and concise aim is just as crucial whether you work alone or in a group. There is consequently reason to seek interaction with others, even if you work alone on your particular project.

A clear and concise formulation of the aim of a study, preferentially in a special paragraph, sometimes even under a special heading, is to many a non-negotiable criterion of a credible text. However, several disciplines do not formulate the aim in such a strict manner. Interestingly, it seems to me as if the disciplines in which the framing is more understood than explicit usually demand a clear and concise formulation of the aim, whereas traditions in which the perspective and other major choices are carefully elaborated upon do not. In the latter studies, the aim or the scope of the study is often understood from the part of the text in which the choice of perspective is justified. Even though the aim is not clearly spelled out, it must still be there, embedded in the text, if it is to be considered credible. Neither Stina nor Peter in the examples above use a specific heading for their aim but both make it very clear for the reader what the aim is, stating "The aim of this thesis is..." (2nd paragraph, Example 9) and "The purpose of the thesis is..." (3rd paragraph, Example 10). Below follows an example from a study in which the aim is understood rather than spelled out.

Example 11
Implicit aim

The heart of Bjørn Lomborg's recent critique of environmentalism is that many assertions of the environmental movement are unproven and therefore provide no good grounds for sensible public policy. Current debate, he argues in *The Skeptical Environmentalist*, is based "more on myth than on truth (Lomborg 2001, p. 32)".1 We all want our views to be based on truth, and many of us look to science to provide truth. But the truth is not always convenient, and it is rarely convenient for everyone, generating incentive for manipulation and misrepresentation of information. This is particularly true in the domain of environmental policy. Lomborg assures us that everyone is for the environment – just as everyone is for world peace and against hunger – but this facile assertion masks the fact that many individuals and institutions, particularly in the industrialized west, have a vested interest in maintaining the *status quo*. Environmental modification is an effect of economic and social activity; preservation, conservation, and mitigation inevitably mean pecuniary or opportunity costs for some individuals, groups, or nations. Demands for intervention engender opposition from those who might expect to bear these costs. Increasingly this opposition takes the form of attacking, impugning, or otherwise seeking to question the science related to the environmental concern (Herrick and Jamieson 2001). In recent years it has become common for informed defenders of the *status quo* to argue that the scientific information pertinent to an environmental claim is uncertain, unreliable, and, fundamentally, unproven. Lack of proof is then used to deny demands for action. But the idea that science ever could provide proof upon which to base policy is a misunderstanding (or misrepresentation) of science, and therefore of the role that science ever could play in policy. In all but the most trivial cases, science does not produce logically indisputable proofs about the natural world. At best it produces a robust consensus based on a process of inquiry that allows for continued scrutiny, re-examination, and revision.

Naomi Oreskes 2004, sid pp. 369–370

The excerpt is from an article published in the journal *Environmental Science and Policy* in which Naomi Oreskes discusses Bjørn Lomborg's book *The Skeptical Environmentalist*. It is understood (but not clearly spelled out) that the aim of the paper is to clarify that the role of science is to provide informed opinions about possible consequences of our actions (or inactions), and to monitor of the effects of our choices (and that it is a misunderstanding that the role of science is to be to a solid proof-provider).

Whether you choose to phrase your aim explicitly (as in Examples 9 and 10) or implicitly (as in Example 11), your work will be easier and more straightforward if you make an effort at the very onset of your study to formulate the aim and/or research questions, and make a habit of always bringing these formulations along when you discuss your work with your supervisor, other scholars or student colleagues. By repeatedly asking yourself "Is this really what I want to achieve?" and encouraging others to question your approach, you will be able to make your study more pointed and increase its credibility.

Operationalizing the aim

Framing a study involves defining the perspective and the aim. On top of that, you must also operationalize the aim, which means that you need to determine what to study to fulfil the aim. Whatever your study is, the aim is to bring a deeper understanding of something from a general point of view. Since you cannot study everything, you have to choose what to study: a case, an example or an experiment, which will illuminate the general problem, phenomenon, system or process. To go from the general to the specific is to operationalize, in other words, to decide how to conduct the specifics of the study. In Example 9, the overarching aim was to study how environmental problems are handled regionally. Stina operationalized the aim by choosing a specific environmental problem (land-based pollution). She further operationalized the aim by choosing a specific region (The Baltic) and even further by choosing two specific organizations (HELCOM and Baltic 21). She could have chosen to study another environmental problem, such as persistent organic pollutants, and she could have chosen to investigate other regional actors such as non-governmental organizations and fishermen. However, she had to narrow down her study, and since the operationalization is not self-evident, it is necessary that the choice of study object be justified. Most research traditions demand that it be clarified and justified in what way a study is operationalized. Some disciplines demand that the operationalization be given in direct conjunction with the aim, as in the case above, in other cases the aim is operationalized in a special section later on. Below follow two examples to illustrate different ways to operationalize the aim, thus choosing the study object.

The overarching question of Natalie Ban's PhD thesis *Multiple perspectives for envisioning marine protected areas*[3] (Example 12) is how to prevent further degradation and to recover degraded ecosystems from the perspective of benefiting biodiversity and people. In her thesis, Natalie explains that she has chosen to focus on marine protected areas (MPAs) in combination with fisheries management (p. 2), as this is the main strategy for marine conservation. She clarifies that the role of humans is a crucial issue, and that the expectation of MPAs is that they will both benefit biodiversity and provide fisheries with benefits (pp. 5, 7–10). In formulating her aim, she narrows down the study and clarifies that the focus is to "enhance our capacity for conservation planning in the marine environment"; she further specifies the focus through five research questions (p. 11). She carefully clarifies how the aim and the research questions have been operationalized, for example by justifying her choice of indigenous community partners on the west coast of Canada as well as her focus on a specific site selection algorithm (Marxan) (p. 12).

Example 12 is from a study that aims at constructing a conceptual hydrological model on catchment level. On a general level, the study consequently deals with construction of a conceptual hydrological model that could be used to predict water movement in any catchment. The author tells us that she has used data from two areas in Sweden while calibrating the model. It is understood that one of the areas has been chosen because it is an international reference site and a second reason is that a year-long data series exists for both sites. However, it is not clear which other sites fit these two criteria and thus were potential study objects and why these, among a presumably larger number of potential sites, were chosen.

Example 12
Operationalizing the aim

Data from two field research areas in Sweden, Velen and Stubbetorp, were used for calibration and validation of the model. The areas are described below. /.../ The Stubbetorp catchment (58 _44'N; 16_21'E) is situated about 120 km SW of Stockholm. The area of the catchment is 0.87 km^2. /.../ The Velen catchment (58 _42'N; 14_19'E) was selected as representative for coniferous forests on till soils during the International Hydrological Decade and after the IHD, it continued to be used as a field research area. /.../

Karin Berg 2003, pp. 5–6

Confusing interdisciplinarity with "Everything"

There are several educational programmes around the world in which the aim is to consciously train students to examine issues from various perspectives, to be familiar with methods and theories of varying disciplinary heritage and to reflect upon the implications of the chosen perspective, method or theory, in other words, to use a holistic approach. Students who have participated in such ventures are often acutely aware that the same question can be answered in numerous ways. My experience is that students may try to find *all* of the answers, since they confuse interdisciplinarity with a naïve understanding of holism. Rather than understanding holism as the whole being more than the sum of its parts, the word seems to be understood in its literal sense, which is "all" or "everything". There is apparently a risk that a "holistic approach" is interpreted as if one is expected to take everything into account, which is impossible and the approach will therefore inevitably result in a shallow study.

One of the key elements of a holistic approach is that no matter how much information you gather, it will never be possible to fully understand or predict the behaviour of a complex system. Various disciplines have developed different ways to demarcate, describe and understand the whole or a piece of the whole. The process by which you formulate the aim of a study, alone or in collaboration with others, helps you to define your choices and thus set the boundaries of your study. In order to study something it is necessary to choose to study some but not all issues, to pick one out of many potential perspectives, to use one or a few out of a number of possible methods – which inevitably means that other issues will not be studied, and other perspectives and methods will not be used. The knowledge gained through the study will be one picture out of several possible ones. Even when you take a holistic approach, a study will *always* only render one description among many of the whole or a piece of the whole.

The methods used in an interdisciplinary study may well be an unorthodox combination, for example of quantitative and qualitative methods (Chapter 8), and the research foundation may well be an unorthodox mix of studies from rather disparate disciplines. Nevertheless, the aim must be clear and concise. Claiming that interdisciplinarity is or must be holistic in its literal meaning is to my understanding to agree with critics who claim that interdisciplinarity by necessity is shallow. Interdisciplinarity neither needs to nor should be shallower than research conducted in traditional academic fields – interdisciplinarity is not a mish-mash of whatever issues, perspectives and methods you might be aware of. It should go without saying that it is neither possible nor desirable to combine all elements from all disciplines. The reason to choose an interdisciplinary approach, to combine or integrate methods, theories, models, concepts or thought styles from different disciplines, is to enable a deeper understanding of the studied field.

Clarifying your choices will help: you may try to write down a plan where you spell out and justify the choices you have made. Writing down and clarifying how you plan to demarcate your study will make it easier to discover in what way the plan needs revision once you have started to implement it. It also makes it easier to ask for input from various people around you: it is far easier to provide feedback on a concrete plan than an idea that is formulated in general terms. A plan that is sufficiently detailed to help you in your revision process is usually more helpful. If it is only a loosely formulated sketch, it will not help you become aware of the choices you make as a result of your revisions. But you should not wait until you have a final, perfect plan before asking for input. All studies change in their making; this is a natural part of the research process.

If you do not write your plan down, it is easy to get distracted from the aim of the study. You believe your plan is clear, it seems clear in your head but when you write it down you will realize that you have made choices that risk leading you to a kaleidoscopic rather than a coherent end-product. This risk is enhanced if you leave the detailed writing to the very last stage, when it often becomes a painstaking process as you will face the challenge of transforming a kaleidoscopic image to a whole. It is usually more fruitful to write an outline at the very beginning of the project and then revise it on a regular basis.

Notes

1 See, for example, Charles Bazermann 1988.
2 See, for example, Karin Knorr-Cetina 1999.
3 Natalie Ban's PhD thesis can be downloaded at https://circle.ubc.ca/handle/2429/1275 or http://en.scientificcommons.org/46175251

8 Evidence that holds for scrutiny

In the previous chapter, I discussed the importance of being aware of the choices you make during the problem formulation process: how to frame a study and how to formulate and operationalize your aim. In addition, for your study to be considered credible, you must convince your reader that you are capable of choosing and using sound and proper techniques that will render researchable information.[1] The sought message is "I know what I am doing and I am able to do it well." Question 7 is designed to help you reach this goal: has the information been collected in a reliable manner and is it of sufficient quality? A complicating factor is that what counts as high-quality evidence varies among disciplines. In fact, it is common that scholars disqualify studies from other traditions because they do not understand what the others do, what they view as evidence and how they go about collecting this evidence. You must therefore be able to explain that you know how to collect evidence in the form of researchable information.[2] Just collecting information you happen to come across is unlikely to render reliable evidence and you must therefore plan the information collection procedure. What type of information should be collected? How will you collect and analyse the information? Do you need to hone and mould the information before you can analyse it? In order to ascertain that the collected information will be of sufficient quality, you must be acquainted with challenges that are tied to the chosen data collection technique; the work will clearly be made easier if you know how the chosen procedure will influence the design.

All procedures demand craftsmanship, and if you are the one collecting the information, you must master the basic techniques. If someone is collecting the information for you, you must know enough about the technique to be able to figure out if the one collecting the information for you has done a reliable job or not.

Interdisciplinary Environmental Studies: A Primer, 1st edition. © Gunilla Öberg.
Published 2011 by Blackwell Publishing Ltd.

As discussed in the previous chapter, designing a study is a question of deciding how to operationalize the aim and thereby identifying which type of information to collect. The information collection procedure also entails storing the information in a way that makes it possible to compile and analyse the material. Posing the following questions holds the potential to help you to not only sharpen your study but also facilitate your understanding of what others do when they collect researchable information, and of what their empirical material is:

- What type of evidence will be/is collected?
- What/who provides the evidence?
- How is the evidence collected?
- How is the evidence stored?
- How is the evidence analysed?[3]

Return to the three papers you chose in Chapter 6, the ones you used to mark theory, method, evidence and analysis. This time, use a marker to identify the answers to the questions above and discuss the outcome with one or several student fellows who have done the same exercise. You can also interview two to three of scholars and ask them to describe what they view as evidence and how they know that it is of good quality. This will help you better understand what in your field is considered compelling evidence and the more people you discuss this issue with, the better you will understand in what way the perception in your field may differ from other fields.

How or why?

To convey the message that the author is capable of collecting reliable evidence, some disciplines focus almost exclusively on *how* a study has been conducted, whereas others focus on *why* it was conducted the way it was. In order to know whether you wish to use a "how" or a "why" approach when you present your study, you need to be acquainted with both traditions and you need to know your audience. Knowing your audience is crucial for all of the issues discussed in this book. In order to make it easier to grasp the difference between "how" and "why" traditions, I describe some key differences and provide one example from each side.

Among "how-traditions", details are emphasized and the descriptions are done in such a way that the study easily can be repeated by anyone who wishes to do so. In the book *How to Write and Publish a Scientific Paper*, Robert A. Day and Barbara Gastel (2006) write:

> In describing the methods of the investigations, you should give (or direct readers to) sufficient details so that a competent worker could repeat the experiments. If your method is new (unpublished), you must provide *all* of the needed detail. If, however, the method has been published in a journal, the literature reference should be given. For a method well known to readers, only the literature reference is needed. For a method with which readers might not be familiar, a few words of description tend to be worth adding, especially if the journal in which the method was described might not be readily accessible.
>
> *Robert A. Day and Barbara Gastel 2006, p. 63*

According to this tradition, the choices are indirectly justified by referring to a standard method (as described above by Day and Gastel), which between the lines means "I have chosen the same procedure as this scholar/ group who have conducted a credible study on a similar topic." Generally speaking, it is in this tradition only in the case of method development that you are expected to justify the various choices made. Below is an excerpt from a rather typical study, which describes the technique for chloride extraction in a cookbook manner, with the choice of technique justified by reference to a (presumably) trusted source:

Example 13
Illustrating "how-tradition"

Chloride was extracted from the ground roots and litter by adding 40 ml 0.1 M $NaNO_3$ to 1.0 g of root/litter and 1.5 g Garco G-60 activated carbon (Gaines *et al.* 1984). Extractions were placed on a shaker table for 48 hours, filtered, and analyzed on a Perstorp Analytical Enviroflow 3500 auto-analyzer.

Shelley J. Kaufmann et al. 2003, p. 25

The recipe ideal, which is so crucial in "how-traditions", is not at all emphasized in the "why-traditions". In contrast to the "how-traditions" the "why-traditions" demand a thorough description of the choice of data collection technique, especially at the undergraduate and graduate level, although this is given less emphasis (if any) in research articles and books. Credibility is here related to how the study as a whole is framed and anchored[4] and the techniques used for data collection are hinted at in the text in a way that signals to the inner circle that this person knows what he or she is doing. In comparison with "how-traditions", more choices are discussed and evaluated in "why-traditions". The first excerpt below is from Stina Karlsson's masters thesis on HELCOM mentioned earlier and exemplifies this. Stina justifies *why* she has chosen qualitative interviews as a research method, justifying the method, the number of interviewees, and the choice to use a Swedish perspective. The second excerpt is from Shannon Hagerman's doctoral thesis on conservation ecology. She is using a form of interview called expert solicitation, which can be used to generate both quantitative and qualitative data. She carefully justifies why she has chosen this method:

Example 14
Illustrating "why-tradition"

Qualitative interviews were chosen as a method to collect the empirical material for this study. The qualitative approach aims at an understanding of the situation of an individual or an organisation by getting close to them /.../ Thus, this research method makes it possible to get an insight into the practice of environmental co-operation in the Baltic Sea region, through the people who work with these issues. The purpose has been to learn about their experiences of and opinions on the work by the organisations of HELCOM and Baltic 21. Consequently, the study gives a view of what the chosen interviewees think and generalisations from the study should only be made carefully.

/.../ To find relevant people to interview, the staff at the secretariats of HELCOM and Baltic 21 was asked for help. /.../ The aim has been to interview people who have insight in issues of diffuse land-based pollution and especially agricultural pollution, within HELCOM and Baltic 21. Persons with corresponding roles and tasks between the two organisations were sought, but difficult to find. Recommendations were given to interview a couple more specific persons, mainly concerned with the policy levels like the creation of Baltic 21, but also people at the Swedish national ministries. However, at that stage the interview material was already so large that with more interviews it would be difficult to do the material justice within this study. On the other hand, if interviewees had been added, the results could have been broadened.

/.../ The study thus takes on a Swedish perspective for practical reasons, like a geographical closeness to the interviewees. This means that a complete picture of the work, including the perspectives of all involved countries, cannot be found through this study.

Stina Karlsson 2003, p. 8

Example 15
Illustrating "why-tradition"

This paper is based on a modified expert elicitation with 21 biodiversity and climate adaptation scientists. Expert elicitation uses structured interviews (or questionnaires) to assess the subjective judgments of experts on technical topics at a given point in time (Morgan and Henrion 1990). The method can yield quantitative or qualitative results and is particularly suited to topics where scientific uncertainty is high, and unlikely to be reduced on a timescale relevant for decision-making (Morgan et al. 2006; Kandlikar et al. 2007). A key strength of expert elicitation with respect to aiding decision-making and identifying future research needs is that it does not seek to identify consensus within a group. Rather, it highlights the current diversity (and locus) of agreement and disagreement within an expert community that may not be voiced in more public fora (Morgan and Keith 1995).

These features implicated an expert elicitation as an appropriate methodology for our study precisely because we were interested in the views of experts (broadly defined as individuals who hold specialized knowledge) on technical topics (e.g. the potential impacts of climate change on patterns and processes of biodiversity) under conditions of irreducible uncertainty. At the same time, given that conservation is seen as "a mission oriented discipline" (Meine et al. 2006) we simultaneously sought to leave open the possibility to examine potential interactions between technical judgment and values in shaping expressed preferences. As a result, our methodology consisted of semi-structured (not structured) interviews with attention to technical concepts as well as expressed value positions, where they were offered. Thus our approach departs from conventional expert elicitations where considerable effort is made to reduce the influence of values, "motivational bias" and other heuristics (Appendix B).

Shannon Hagerman 2009, p. 122

Common procedures

Students in the humanities and social sciences most often take numerous courses on general research methodology, whereas such courses are less common in the natural sciences and engineering. Even though many of the questions dealt with in the literature on general research methodology are of relevance for all empirical studies, the literature seldom speaks about similarities across scholarly traditions. On the contrary, this literature often emphasizes that the issues dealt with are only valid in the humanities and social science sphere. Hence, even if students in the humanities and social sciences have a thorough training with regards to general research methods, they have little or no training in identifying similarities and common ground across research traditions.

The aim of this section is to elucidate such similarities. I assume that some of my readers are unfamiliar with literature on research methodology, whereas others are well read in the field. I try to use a language that can be understood by the former while discussing issues that are of relevance for both groups. Through this brief synthesis, I hope to encourage cross-cutting studies.

Interviews, experiments, observation and document studies are common procedures in empirical research across the humanities, social sciences, natural sciences and technology. There are numerous books that delve into details regarding these procedures. The following is *not* a how-to-guide on how to conduct these kinds of studies. The aim is to compare four types of research procedures to illuminate methodological challenges that are common to most empirical studies. The focus is on study design and data collection procedures and I discuss the other three procedures in light of interviews.

In the following, I make a crude simplification by speaking of "the social sciences" on the one hand and "the natural sciences" on the other hand – hoping that those who are rooted in the humanities, technology, health sciences, etc. understand that they are implicitly included in the two crude categories I use.

Interviews

Interviews are used in a number of research traditions. The *questions* determine whether it is a natural science study (questions focusing on processes and phenomena in Nature) or a social science study (questions focusing on processes and phenomena in society). Surveys can be seen as a form of interview where the researcher interacts indirectly with one or several persons.

Study design
An interview is usually carried out as a conversation with one or several persons. The conversation is conducted in such a manner that it makes it possible to use the answers as a source of researchable information. To identify which type of information you will collect is a question of formulating the questions you will pose to the person or persons you are interviewing. It is crucial that you are aware of the difference between an ordinary conversation and an interview or professional conversation. The

structure of the interview determines to what extent the interviewer permits the development of a conversation. The least open and strictest form, the structured interview, can be carried out as a conversation with closed questions and given response alternatives (e.g. on a scale from 1–10 how would you ...). A slightly more open form is the use of closed questions with open response alternatives: Question 1 – answer – pre-prepared question 2 – answer, etc. In the semi-structured interview, the interviewer prepares a questionnaire, just as in the structured interview, but the form of the interview is different. The semi-structured interview strives to mimic everyday conversations in which the interviewer introduces the interviewee to the subject field, whereupon the conversation is allowed to develop within certain limits; the questionnaire is used as a conversation support, making it possible for the interviewer to check that all questions are covered during the conversation, irrespective of the order or length of the answer. The least structured form is the open interview, in which the conversation is allowed to develop freely after an introduction.

Data collection
The data obtained through an interview may be transformed to research-able information in mainly three ways: i) recording/videotaping; ii) notes taking during the conversation; or iii) notes taken after the conversation has been carried out. Recorded and videotaped interviews are transferred to notes. The format of the written notes vary largely, from a summary synthesizing the subjects discussed to elaborated transcriptions including sounds, pauses, intonations and descriptions of body language.

Observation

Observation is a method in which the researcher observes something (an event, a development, a process) continuously or during intervals within a certain period of time. It may be a short-term event (such as a boardroom meeting or two rabbits mating), or a temporally-extended process (such as changes in traffic patterns or planetary orbits). Observations can be carried out in all research domains and they may be qualitative or quantitative. Participatory observation is a type of observational studies, introduced in anthropological studies in the early 20th century. The method is also widely used today within ethnography and sociology.

Study design
When designing the study, you have to decide what to observe, how to conduct the observations, and how to store the information. On an over-

arching level, observations can be viewed as a type of interview: the researcher prepares questions. In this kind of "interview", however, it is either not possible to ask direct questions (as with the rabbits), or direct interaction with the object of study is likely to skew the results. Instead the researcher records and studies things that are happening, to find the response to the posed questions. When conducting an observation study, you must be aware of the difference between everyday observation and observation that will render researchable material. Everyday observation, for example, may be a question of noticing that the refrigerator stops making noises when you step on the threshold or that the snow last week was the worst in your life. Scholarly observation must be directed such that the information which is collected is of such quality that it can stand up to rigorous analysis.

In participatory observation, the researcher spends time with and inter-acts closely with the person or group of people about whom he or she conducts research, in order to acquire a deep understanding of the thoughts and life of that person or group. Field notes are usually taken and used as a basis of the study. The empirical material is thus partly in the head of the researcher and partly found in the field notes. Since no one can control whether or not the story told by the researcher is reliable, descriptions are generally elaborated upon in great detail to make it possible for the reader to evaluate the study's credibility.

Data collection
When observation is used as the method for collecting researchable mate-rial, the researcher generally keeps a logbook and takes field notes: these self-produced texts, in combination with the researcher's memory of the studied context, comprise the whole or part of the empirical material. In addition to the field notes, researchers in some disciplines use some sort of instrumentation to observe the studied phenomena, and the observations are recorded in one way or another. The information that has been recorded (in combination with the field notes) is the researchable information. Whether you observe humans, things, processes in society, processes in Nature or technical systems, the procedure is rather similar. When observing things, for example, the researcher may observe physical phenomena such as sound, light or currents; chemical phenomena such as the concentration of certain chemical substances, redox potential or transformation rates; biological phenomena such as the number of a certain organism or changes in species composition; or technical phenomena such as dams, communica-tion systems or engines.

The first example below is from Karin Wesström's Honours thesis, which aims at elucidating how organic chlorine is distributed in soil. Karin describes the information-gathering procedure in detail. She describes the procedure used to observe certain chemical characteristics of the soil (collect soil samples, bring them to the laboratory, conduct chemical analyses and conduct statistical data analyses). The second example is from Shannon Hagerman's doctoral thesis in which she uses a combination of participant observation, observation and interviews:

Example 16
Illustrating data collection through observation

The catchment area was divided into a grid with eight lines in north-south direction and eight lines in west-east direction located approximately 105 m apart. Soil samples were collected on April 5–7, 2002 at 49 of the nodes on the grid. /.../ The samples were stored in plastic bags at −10°C for up to two weeks before analysis. /.../ The total amount of organically bound chlorine in the soil samples were analysed according to the TOX method /.../ The chloride content was determined by /.../ Since the studied data material is not normally distributed, the central tendency is given as median and the variation is presented as 1st and 3rd quartile

Karin Wesström 2002, p. 7

Example 17
Illustrating data collection through
observation and interviews

Specifically the analysis presented here is based on: 1) detailed participant observation at more than 13 workshops, knowledge café's and Pavilion events during the Forum portion of the Congress; 2) observations conducted during Contact Groups and Plenary sessions of the Members Assembly; and 3) 8 semi-structured interviews (between 45 minutes to 90 minutes) with leading biodiversity-climate change experts from domains including academe, NGOs and from the IUCN secretariat. Combined, these research activities amount to over 50 hours of observational data on the content and nature of debates around the implications of climate change for biodiversity conservation. The strength of this approach as applied in this policy-making setting and in relation to our questions is to reveal nuance between the perspectives of individuals voiced in private, and how and why these perspectives are mobilized (or not) in social contexts where specific objectives are sought.

Shannon Hagerman 2009, p. 159

Experiments

Various types of experiments are commonly used in a number of disciplines. When conducting an experiment, the researcher consciously interacts with the study object in order to provoke a response. In an experiment, you interact with things, humans or a technical system in such a manner that it makes it possible to use the response as a source of researchable information. It can be a question of investigating how a forest responds to liming, how a group of consumers responds to a certain type of advertisement or how the methane formation in a dam is influenced by a fluctuating water level. Experiments can be conducted in all research domains, but must be ethically approved if the study deals with humans or higher animals. Experiments may be qualitative or quantitative, just like interviews and observation studies.

Study design

Conducting an experiment means that you manipulate something in a way such that it is possible to use the response as a source of researchable information. To design the study is a question of determining what to manipulate and in what way. In part this is a question of what type of response you will observe, how these observations will be conducted and how the information from the experiments will be stored. Experiments can be viewed as a form of interview: you "ask a question" by provoking a response, and the response is your "answer". Speaking in interview terms, experiments can be seen as structured, semi-structured or open interviews. The strictest form is the structured experiment/interview in which all variables but one are controlled and the response can thus conclusively be tied to that variable. An open experiment is more a question of disturbing a system without really knowing what variables might react, why you observe a wide spectrum of variables. Whether or not the experiment is strictly structured or open, it is not possible to directly record the response. Instead the response must be observed, and the observations must be interpreted. When conducting an experiment, you must be aware of the difference between everyday experimenting and experiments that have been designed to deliver researchable information. An everyday experiment may be to test various ways to make the refrigerator noise stop and to find out that punching the left upper corner does the trick.

"Natural experiments" is a phenomenon or process outside the ordinary. For example, the 2004 Indian Ocean tsunami has been used as a natural experiment in which adaptation to natural disasters by societal and natural systems have been studied.

Data collection

The gathering of information when conducting an experiment is identical to the procedures used in observation studies. Below I exemplify with an excerpt from Anna Lundén's masters thesis in which she conducted a laboratory study with a bioreactor. She collected sludge samples from the reactor as well as air samples above the sludge. Temperature, humidity and oxygen levels were controlled, but in other respects the system was allowed to develop freely, and it can thus be described as a semi-structured experiment. Anna describes which variables she analysed and the results are presented in figures and tables:

Example 18
Illustrating data collection through experimentation

Three biogas reactors have been used: one reactor in which the grains were diluted in tap water (TW), one with reject water (RW) and one with sludge water (SW)./.../ 100 ml of sludge was removed from each reactor daily in order to get a 30-day retention period and maintain a constant volume in the reactor. From these 100 ml, samples were allocated and used for different analyses. A sampling plan was constructed and adhered to as closely as possible, though there were of course variations due to practical obstacles. /.../ In this study, there were variables that were monitored on a daily basis and others that were monitored on a weekly basis or a couple of times a week. Some variables need to be monitored daily as the information provided by the results are valuable indications as to the current status of the reactors. The variables monitored on a regular basis, but not daily, were useful complements to information obtained daily as well as providing a way to compare results with standard nomenclature. /.../

Anna Lundén 2003, pp. 13–15

Document studies

Information found in various types of documents, such as books, journals, newspapers and research articles is used in numerous disciplines as a source of researchable information. If you use historical documents, for example, it may be a question of finding out how people's eating habits have changed by analysing cookbooks, diaries and annals from different epochs. It may be a question of finding out how the weather and climate have changed by studying temperature logs, weather reports and diaries. It may be a question of investigating how our view of masculinity has changed by studying newspapers and other relevant documents. Document studies can be conducted in all research domains and they may be quantitative or qualitative.

Study design

Document studies, or textual analyses, are collective names for studies in which you analyse the content and meaning of a text. Such studies are sometimes confused with comprehensive reviews. The author of a comprehensive review tries to make an objective description of a certain field; the aim is not to create new knowledge but to summarize or synthesize contemporary knowledge within a field.[5] All types of empirically based research demands that empirical material is analysed/evaluated/scrutinized in light of previous research. A review is a summary/compilation of other studies: the information presented is referred to rather than analysed, so a review can therefore not be said to be an empirically based study. A prerequisite for a good review is that the author is well acquainted with the field. In contrast, the aim of a document study, a critical review or a textual analysis is to create new knowledge. The author analyses information that has been systematically collected from the text/s. One approach is to describe and analyse information that is explicitly stated in the text. Another approach is to analyse implicit meanings.

All good studies demand that the author collects relevant information in light of the aim of the study. Whether you conduct an interview, send out a survey, or observe or experiment with things in Nature or society, it is crucial that the sources of information have been strategically chosen. The same afterthought is necessary when conducting a document study – the author must be able to justify why certain texts have been chosen for the specific study. When conducting document studies, you must be aware of the difference between everyday reading and reading a text with the intention of collecting researchable information.

Data collection

A document study can also be seen as an interview. Whether you conduct an interview or a document study, the researcher (implicitly or explicitly) prepares a questionnaire. One might say that the researcher interviews the text but, since the text cannot respond, the researcher must scrutinize the text to find the answers. Some traditions demand that the questions posed of the text are explicit, while others formulate their questions in the process of reading the text.

Analyses of the verbal content of a document have many similarities with analyses of interviews and surveys with open questions. As mentioned above, the empirical material resulting from an interview is generally either notes or a transcription, i.e. a written document. The methods used to analyse texts originating from an interview are very similar to the methods used to analyse any other type of written document. The major task in both cases is to interpret something "said" by a human. Hence, empirical material that is based on excerpts from a document has many similarities with empirical material collected through interviews. The major difference is that the interview originates from a conversation in which the researcher has participated, and the interpretation of the notes/excerpts from an interview is thus coloured by impressions from the interview situation.

Below follows a citation from Emma Larsson's master's thesis in which she elaborates upon the data-collection procedure of a document study and the problems related to this. The questions posed to the texts in this example are clearly formulated in the aim; in addition, the analytical categories are formed prior to the reading. The second example is from one of the papers in Shannon Hagerman's doctoral thesis in which she clarifies which questions she has been asking while reading "historical documents relating to forest policy and conservation, international conventions and the scientific literature on forest disturbance and uncertainties in this system:"

Example 19
Illustrating data collection through document studies

/.../ To approach the aim, I have taken my point of departure in the extensive written record of the IPCC and the FCCC. Documents reflect conscious and unconscious conceptions of the people that are involved in the creation of the texts, and can therefore say something about these institutions' perception of the climate change issue.[54] A document can be regarded as a stratification of social practices with a potential of having an affect on both short-term and long-term decisions.[55] However, documents cannot be regarded as reflection of reality since they are shaped in a specific context and affected by social and cultural commitments. One of the greatest benefits with using texts as an empirical basis is their stability. In contrast to observations and interviews the researcher does not influence or change the object in focus of the study by his or her mere presence.[56] /.../

The hermeneutic process is dependent on the researcher's own power of initiative to start a dialogue with the text.[64] In order to start a dialogue a number of questions were formulated and asked to the texts, in a similar way that can be done in an interview. Through the

Continued

process of analysis it is important to consider how the context affects the production, consumption and interpretation of the text.[65] The applied method is very similar to what Bryman (2002) describes as qualitative content analysis. This method includes the search for underlying themes in the material, which also was in focus in this analysis. In the analysis both the implicit and explicit aspects of the texts were searched for, since the actual meaning can be difficult to find.[66] I have been trying to elucidate and conceptualize the essential content in the texts in relation to the aim of the thesis and the research questions defined below.[67]

Emma Larsson 2004, p. 6

Example 20
Illustrating data collection through document studies

This paper represents an effort towards an integrated understanding of the biophysical and socio-political drivers operating in BC forests over time and across scales. We use the provincial boundary to delineate the system because this is the scale of the major forestry and conservation legislation, and because it serves our purpose of drawing coarse scale insights into the dynamics of change in this region. Critically, we also examine drivers external to this boundary. In order to explore this integration, we asked the following questions of historical documents relating to forest policy and conservation, international conventions and the scientific literature on forest disturbance and uncertainties in this system:

- How has a selected set of system attributes changed over time?

- What drivers triggered change when it occurred? From what scale and what domain did these triggers originate?

- What do the historical change dynamics suggest about further iterations of change in this system?

In asking these questions we simultaneously explore 4 propositions relating to policy change more broadly: (1) the degree of influence of different drivers varies with origin (exogenous vs endogenous to the system), domain (e.g. social, ecological), and over time; (2) the presence of decision-relevant uncertainties is not, in-and-of-itself, a barrier to the adoption of new policy proposals; (3) policy change in this system has proceeded according to a punctuated equilibrium model; and (4) assumptions of ecosystem stability persist in the latest policy iterations.

Shannon Hagerman 2009, p. 96

Mixing various types of empirical evidence

One of the strengths of interdisciplinary studies is that it arranges for fruitful combinations of methods, theories and practices from different disciplines. Natalie Ban, Shannon Hagerman, Jane Lister and Charlie Wilson all used different combinations of interviews, observation and document studies in their doctoral theses and are excellent examples of how methods from different academic traditions can be mixed. When elements from various disciplines are combined consciously and with rigour, the research has the potential to break new ground and bring about considerable leaps in your ability to explain and understand various phenomena and processes. Even if uninformed and random fusions of disciplines may in some cases lead to new insights, this far more often results in a mess, even if a relatively harmless mess; and in worst case scenarios, it amplifies misunderstandings, enables an unhealthy politicization of research results and damages the reputation of research in general and interdisciplinary research in particular. Hence, I am strongly in favour of research that combines methods, theories and practices from various disciplines; at the same time, I stress the need to take precautionary steps to avoid charlatanry. The exercises outlined in the beginning of this chapter will help you become more aware of how quality demands vary among academic traditions, which in turn will help you design a rigorous study that combines different approaches.

Notes

1   In other words, information collected in such a manner that it can be used in research. Different disciplines use different terms to denote this type of information. Common terms are, for example, "data," "empirical material" "information," and evidence but all of these terms have different connotations in different contexts; to avoid misunderstanding, I have chosen the descriptive term "researchable information," which later is used interchangeably with "data" and "empirical material."
2   The procedure used to collect researchable information is by some disciplines called "method," whereas others use the term to denote the technique used to collect the information; yet others use the term "method" to denote the whole research procedure. I use the term to denote "the procedure used to collect researchable information".
3   The term "analyse" can be used with at least three different meanings, which may complicate interdisciplinary collaboration. More about this issue in Chapter 10.
4   More on anchoring in Chapter 9.
5   Although a crucial issue, it is beyond the aim of this book to further elaborate upon whether or not it is possible to make a matter-of-fact, value free and neutral description of a research field and, if so, whether this may be done without creating new knowledge (see, for example, Steinar Kvale 1996 and Ludwik Fleck 1979).

9 Anchoring your canoe

When you write an academic text you must convince the reader that you know your field. The difference between academic texts and other writings is that the former must be securely anchored in previous academic work. The author is expected to explain how the work relates to earlier and contemporary studies and what it brings to ongoing scholarly debates and discussions. Your text can be seen as a conversation between you and the reader, where you continuously bring others into the dialogue. You lend other authors your voice, you interpret what they have said, thought and done and you relate this to your own work and thoughts:

> Scientific citation is the process by which conclusions of previous scientists are used to justify experimental procedures, apparatus, goals or theses. Typically such citations establish the general framework of influences and the mindset of research, and especially as "part of what science" it is.
>
> *"Scientific citation" according to Wikipedia, February 8, 2008*

Anchoring your work means that you clarify when you refer to findings originating in your study in contrast to when you make references to ideas, data, methods and results that originate from other sources. Even if you lend your voice to previous writers, you must clarify when you speak on someone else's behalf and when you speak on your own behalf. Question 8 is designed to help you achieve this end: Is your study sufficiently anchored in relevant literature? To speed up your understanding of the anchoring process, you can use the same papers that you used for the exercise in Chapter 6. As a first step, use two markers of different colours to identify when you refer to findings originating in your study in contrast to when you make references to ideas, data, methods and results that originate from other sources. Meet with a group of fellow students who

---

Interdisciplinary Environmental Studies: A Primer, 1st edition. © Gunilla Öberg.
Published 2011 by Blackwell Publishing Ltd.

have done the same exercise and discuss your markings. This will increase your ability, in a reliable way, to anchor your study in previous work, it will help you better understand of how the requirement to anchor your text varies among traditions and it will help you find common traits.

It is commonly agreed in academia that a study must be carefully anchored in relevant literature and the origin of the information must be clearly presented. You have probably already discovered that these two common demands are hidden in disciplinary differences regarding what to be written where and how.[1] In this chapter, I discuss how the structure and style of the anchoring process may cause problems in interdisciplinary environments and how such barriers may be overcome. I start with some comments on the importance of clarifying your sources.

Clarifying your sources

> In winter at that time there was a local study circle on psychology led by Mr Bergqvist, the schoolteacher in Glommersträsk, and there was a report on it, as there was on everything else, in the newspaper. The study guide was written by Dr Alf Ahlberg. The life of the soul was, according to Plato, as Mr Berqvist said Dr Ahlberg explained, the only real and true life.
>
> *Torgny Lindgren 2002, p. 132*

The excerpt above is from the book *Hash* by Torgny Lindgren, which is a wonderful Swedish dark comedy that hardly can be called an academic piece of work. The excerpt is nevertheless most illustrative of how you may go about clarifying the origin of information. The cascade of sources in the last sentence is no doubt very clear, although one might want to add "as stated in the newspaper, according to the narrator in the novel *Hash* written by Torgny Lindgren."

In contrast to novels, where the author is free to mix fiction and truth to his or her liking, you are, as an academic writer, expected to truthfully clarify from where statements, comments, facts and conclusions originate. It is an absolute must that you clarify which statements are the results of your study and which statements originate from someone else. You must learn how to clarify the origin of statements, comments, facts and conclusions and, for reasons outlined further on in this chapter, this is particularly important when conducting interdisciplinary work.

In most undergraduate programmes, it is stressed that students must learn how to distinguish between citations, accounts, plagiarism and their own conclusions. In the textbook *How to Write and Publish a Scientific Paper*, widely used in undergraduate science teaching, Robert A. Day and Barbara Gastel (2006) write:

> Good scientists build on each other's work. They do not, however, take credit for others' work. If your paper includes information or ideas that are not your own, be sure to cite the source. Likewise, if you use others' wording, remember to place it in quotation marks (or to indent it, if the quoted material is long) and provide a reference. Otherwise you will be guilty of plagiarism.
>
> *Day and Gastel, 2006, p, 27*

A student who has reached the master's level is, in most disciplines, expected to be well acquainted with the distinctions between citations, accounts, plagiarism and his or her own conclusions, and it is expected to be a conditioned reflex at the post-doctoral level. In spite of these demands, students recurrently fail because they have been citing in a misleading manner, they have neglected to refer to a source or they have engaged in deliberate plagiarism. Most of the time, it is more a question of ignorance than conscious cheating. Citing in a misleading manner, however, is academic misconduct. Researchers who take to plagiarism or who cannot account for their sources will soon lose credibility. If you want people to view your work as trustworthy, you must learn the art of citing.

Below follow two sections – one on framing and one on methodology – which each illustrate different ways to anchor academic work. Anchoring of the analysis is discussed separately in Chapter 10.

Anchoring your frame

As mentioned in the previous section, academic writing must be contextualized. The following example may be a bit profane, but the first part of an academic text is somewhat similar to the introduction of a detective story in which the author familiarizes the reader with the settings in which the crime is to take place. If the reader is introduced to a provincial village in Scotland and a murder then takes place in Viet Nam, the reader will spend the rest of the reading wondering about the connection to Scotland, and if there is none, the reader will most likely become quite irritated. The same goes for academic texts: if the start does not introduce the reader to the subject in such a manner that it is clear why certain steps are taken and how different parts are related, the study will not pass as credible. In

Chapter 7, I discussed how framing a study involves choosing a perspective. In addition, the process of framing entails choosing a canon; a body of literature which will be used to contextualize the study. The aim of a study is generally reformulated over and over again as a natural part of the research process. In the process of reformulating the aim you must also revisit the framing and thus also the canon. The Scottish village might have been a relevant part of the original story, but if that part of the story is cut out – then it should also be cut out of the framing.

The framing is crucial for all academic texts, but the language, the terminology and not least of all the length of the text used to frame the story varies with disciplinary traditions and the subject. Some disciplines demand a chapter called Background, Theory or Theoretical Framework. These chapters are sometimes textbook style reviews of literature of relevance in a rather wide sense. One could say that the work in these disciplines is guided by the idea that "more is beautiful". Below I give an example as an illustration. It is a study with the aim of elucidating whether illegal logging and its consequences may lead to conflicts in Indonesia. The author uses three complementary theories and she gives a rather thorough background description of these theories plus a background description of Indonesia. All in all, the background descriptions in the paper comprise two-thirds of the thesis.

Example 21
Illustrating one way to anchor the frame

1.1 Aim of the study

The aim of this study revolves around three different conflict theories, which will be used to analyze the current impact of the illegal forestry business in Indonesia. The main purpose is to investigate if and how illegal logging and its causes and consequences in Indonesia could lead to conflicts. Necessary questions to be answered in regards to the aim of the study are:

- What are the causes of the illegal logging?
- What part does corruption play in the illegal forestry business?
- What are the consequences of the illegal logging for the environment and people dependent on the forests?

/.../.

Continued

> 3.   Theories on environmental conflict
>
> To be able to answer the problem formulation, three different theories around the topic environmental conflicts have been reviewed. The following chapter is aimed to give the reader a short introduction to their contributions which lays the foundation for this thesis. The theories reviewed in this paper are closely linked to each other, but still portray different perspectives which rather than overlap, compliment each other, and thus give a nuanced picture of the researched area
>
> *Sofia Weiss 2003, pp. 7 and 11*

Other disciplines confine the framing to a summary of key literature in the introduction and allow only references to literature that is of *direct* relevance to the study, guided by the idea that "less is beautiful". Those adhering to this view of academic writing would most likely demand that the introductory parts in Example 21 be made considerably shorter, only keeping the parts of direct relevance to conflicts related to illegal logging, and that the author, in a short summary, simply refer to major works on the theories on environmental conflicts for the interested reader.

The discussion in the example above serves to illustrate that there are different ideas on how much of the canon is expected to be presented in the text. It is hardly surprising that the framing often is a source of dispute in interdisciplinary contexts, especially when it comes to judging jointly supervised students. Some supervisors deem a paper to be of insufficient quality if it does not contain a thorough and rather broad theory chapter, whereas others deem a paper to be of insufficient quality if the literature referred to is not of direct relevance to the study. I argue that there are no reason to demand rigidly that all papers have the same form and structure. In an interdisciplinary environment, you must be willing to accept a certain degree of flexibility. One of the challenges you face as a student is that some of the scholars you meet will be unaware of the "beauty comes in many forms" principle. You must therefore find a way to be diplomatic and humble as well as firm. Diplomacy is needed when you clarify why you have chosen a certain style. Humbleness is needed since you must stay open to the possibility that there is a better way of doing things than the way you have chosen. The more senior scholars around you are undeniably more experienced than you and you can learn a lot from them. There is no reason to be arrogant, but you need to be firm. The argument that "this is how it is supposed to be done" is never a valid argument in an interdisciplinary context. Demand (humbly and diplomatically) an explanation why a particular style works best for your work and change paths when you are being genuinely convinced rather than browbeaten. But, but, but – don't be stubborn. Review your own arguments – are you perhaps clinging to a "this is how it is supposed to be done" idea?

If you wish to speak to a wider audience, it might be wise to use a style that can be understood and accepted by many. One possibility is to use a critical analytical approach to review key literature in the introduction. A critical review will be considerably longer than introductions of texts in traditions that expect condensed summaries but considerably shorter than theory chapters in traditions that demand textbook style comprehensive reviews. Using a critical analytical approach, will help you clarify for the reader (and yourself) in what way the reviewed literature is relevant in light of the aim of the study.[2] This will make it easier for those expecting both shorter and longer texts to accept your approach.

Anchoring your method

As discussed in Chapter 8, you must convince the reader that you know how to collect researchable information, which means that you must carefully explain what you have done. This is in part a question of demonstrating how your way of doing things relates to and builds upon other researchers' way of doing things. You are expected not only to clarify what kind of information will be collected and how; how the information will be analysed; what theoretical basis the method draws upon, and so on. You must also convince the reader that the study has been conducted in a solid, rigorous and acceptable manner. The procedure for collecting empirical material, the method, must therefore be anchored in previous research.

Different disciplines have different ways of doing this. I suggest that you use the same papers that you used in Chapter 6 to identify how they anchored their method. Compare notes with your student fellows: how have the authors of their favourite papers handled method anchoring?

Below I provide four examples and discuss the way the authors have chosen to anchor their methods. Sometimes, the descriptions tend to be rather technical, studded with abbreviations and acronyms, as in the example below.

Example 22
Anchoring the Method by referring to a reliable source

Sampling, Preparation, and Characterization of Soil Organic Matter Fractions. A 30 cm thick organic horizon from a Gleysol (13) was sampled at Svartberget Research Station, Vindeln, Sweden. A SOM (soil organic matter) sample was homogenized and separated from plant roots and large undecomposed debris using a 4 mm cutting sieve. Total organic carbon was 46% and total nitrogen 2.2% of dry mass, as determined by dry combustion (2400 CHN elemental analyzer, Perkin-Elmer, CT). Soil pH was 3.16 determined at a 1:6 soil to solution mass ration in 0.01 $MCaCl_2$. Adsorbed metal ions were sequentially extracted with 0.5 M $CuCl_2$ using the method of ref 14 and were determined on ICP AES (Perkin-Elmer). The concentration of adsorbed Na, Mg, Ca and Al was 1, 3.5, 10 and 115 $mmol\,kg^{-1}$, respectively. The total cation-exchange capacity (CECt), calculated as the sum of total acidity determined at pH 8.2 (15) and the sum of charges pertaining to adsorbed Na, Mg and Ca was 2110 $mmol\,kg^{-1}$ soil. /.../

Literature Cited

/.../

(13) Soil Survey Staff. SMSS Technical Monograph 19. Pocahontas Press: Blacksburg, VA 1992.

(14) Skyllberg, U.; Borggaard, O. K. *Geochim. Cosmishim. Acta* 1988, **62**, 1677–1689.

(15) Thomas, G. W. In *Methods of soil analysis, Part 2. Chemical and microbiological properties*, 2nd edn: Agronomy Monograph 9: Page, A. L., Miller, R. H., Keeney, D. R., eds, ASA-SSSA: Madison, MI, 1982: 159–165.

Johan Eriksson et al. 2004, p. 3075

In the excerpt above, we are informed that a copper chloride solution has been used to extract metal ions "using the method of ref 14." For this type of research, the type of instrumentation has a strong influence on the type of information that may be obtained. The authors therefore not only mention the type of instrumentation but also the brand, as when they write "(2400 CHN elemental analyser, Perkin-Elmer, CT)." In plain English, the information in the parentheses tells us that the authors have used an instrument that is able to detect carbon, hydrogen and nitrogen by converting it to its elemental form and that the instrument is produced by the company

Perkin-Elmer. The procedure is clearly rooted in the literature and the techniques that have been used can easily be evaluated by the reader. If the reader is unfamiliar with the techniques, they can easily be looked up through the references and combined with information on the limitations of the instruments, which usually is given on the home page of the firms who produce the instruments. In this type of study, the authors convey reliability and rigor by clarifying that they have used techniques that other researchers have used previously with reliable results.

The authors do not justify the choice of technique, the instrumentation or the brand. This is fully in line with the tradition, as it is considered sufficient to refer to a previous study without further justifications. The crucial part is that the description is made in such a manner that it is possible to repeat the study. To reflect further upon pros and cons is considered superfluous.

In other disciplines, merely referring to relevant literature is not sufficient: as the method is described, you are also expected to reflect upon the issues that need to be dealt with. Example 23 is a study in which the authors develop a technique for distinguishing between naturally and anthropogenically formed halogenated substances. The technique draws upon a number of previous and similar but not identical approaches, and the authors clarify what this new technique is based upon and what the strengths are; later in the discussion, they also touch upon potential weaknesses:

Example 23
Anchoring the method by discussing previous studies

Organochlorine (OCl) substances are of wide environmental interest due to their generally high persistence, bioaccumulation potential, and ecotoxicity. The chlorine substituents will to a large extent determine the molecular properties during transport processes and chemical reactions. Hence, transformation and transport of most OCls impinge on their compound-specific chlorine isotope ratios. These shifts in isotope ratios reflect the types and extents of processes experienced by the OCls. Compound-specific isotope analysis (CSIA) of stable chlorine isotopes has the capacity to provide knowledge relevant to a wide range of environmental[1,2] and geochemical[3] issues. Degradation rates, distribution patterns, and reaction pathways for OCls may be elucidated with CSIA. Apportionment of sources of OCls by CSIA is a strong potential application in environmental[4] and forensic science.

Continued

Isotope ratios of stable chlorine are expressed in $\delta^{37}Cl$ notation according to:

$$\delta^{37}Cl_{SMOC}) [R_{sample}/R_{SMOC} -1]\times1000 \qquad (1)$$

where R signifies the ratio of $^{37}Cl/^{35}Cl$ in sample and reference material, respectively. The result is the per-mil deviation in $^{37}Cl/^{35}Cl$ of the sample vs the reference, usually Standard Mean Ocean Chlorine (SMOC). Chlorine isotopes are believed to exhibit homogeneous abundance throughout the oceans, due to the relatively short circulation time of the oceans (1.5×10^3 y), as compared to the long oceanic residence time for chlorine (87×10^6 y). The absolute chlorine isotope ratio of seawater has been determined to be $^{37}Cl/^{35}Cl$ 0.319644/0.000917 (IAEA standard ISL354)[5] by relating to the absolute value of NIST standard SRM975 (solid NaCl). Most applications of chlorine isotope analysis require significantly smaller uncertainty than that associated with the ISL354 absolute value (2.9‰).

The entire procedure of CSIA consists of extraction and isolation of individual compounds from complex mixtures,[6-7] followed by isotope analysis of the compounds. Isotope analysis is performed either off-line (e.g. $^{37}Cl/^{35}Cl$, $^{14}C/^{12}C$) or through continuous flow (on-line) from the gas chromatograph to the mass spectrometer (e.g. D/H, $^{13}C/^{12}C$, $^{15}N/^{14}N$). The present work describes a method for chlorine isotope analysis of OCls which reduces the amount of compound that has to be isolated.

Henry Holmstrand et al. 2004, p. 2336

Scholars in the social sciences and humanities usually demand that students be able to reflect upon their choices at the undergraduate level within previous method literature as their point of departure. However, in comparison with the natural sciences, less emphasis is placed on the details of the technique used. In one of the papers in her doctoral thesis, Shannon Hagerman first describes the interview technique by reference to an established source:

> /.../ Expert elicitation uses structured interviews (or questionnaires) to assess the subjective judgments of experts on technical topics at a given point in time (Morgan and Henrion 1990) /.../
>
> *Shannon Hagerman 2009, p. 112*

Thereafter, she justifies the choice, again with reference to an established source;

[...] The method can yield quantitative or qualitative results and is particularly suited to topics where scientific uncertainty is high, and unlikely to be reduced on a timescale relevant for decision-making (e.g. Morgan *et al.* 2006; Kandlikar *et al.* 2007).*[...]*

ibid, p. 112

Shannon thereafter moves back and forth between the general – how the methods are described in the literature and the specific – how she has implemented the methods in her study, thereby carefully rooting her study in previous method literature.

Notes

1 To become familiar with differences in style among different academic traditions I recommend writings by Janet Giltrow, for example, *Academic Writing: Writing and reading across the disciplines* (3rd edn, 2002).
2 I discuss the challenge of identifying relevant literature in the section Relevant literature – your canon.

10 Analysis

To evaluate, interpret and discuss research findings in light of previous work is perhaps the most challenging part of academic scholarship – challenging to teach and challenging to learn. Question 9 is designed to help you achieve this goal: Is the study analysed in an informed and reflective way? I suggest you return to the three papers you picked for the exercise in Chapter 6, where you used four markers of different colour to identify text describing theory, method, evidence and analysis. Scrutinize the sections you and your friends have marked as "analysis", and initiate a more thorough discussion on the meaning of "analysis" and what a good analysis is. This may help you avoid the common newcomer mistake of writing a non-analytical front-loaded paper: a long and detailed textbook-style introduction followed by a brief presentation of the aim and design of the study and a condensed summary of results, and then the paper ends abruptly without any trace of an analysis. To write an informed and reflective analysis takes practice, and you are more likely to succeed if you practise along the way rather than taking a first stab at the end of your project.

Another way to approach the task is to start practising at the very onset of your project by reviewing the literature in light of your aim and review your aim in light of the literature – in plain written language. You do not have to write long paragraphs. You can make short summaries of your readings and comment upon them in your own voice. By constantly distinguishing your own thoughts on the literature from thoughts, facts, conclusions and the like in the literature, you will, among other things, minimize the risk of unconscious plagiarism (see Chapter 9). If you make it a habit to keep written reflective notes on the literature, you will learn how to relate to it and, when your results start to unfold, you will find it much easier to discuss your findings in light of the literature. Search for a style that suits you and that makes it possible to clarify the roots of the thoughts you formulate.[1]

Interdisciplinary Environmental Studies: A Primer, 1st edition. © Gunilla Öberg.
Published 2011 by Blackwell Publishing Ltd.

Some teachers picture a staged learning process where the first stage is to acquire the ability to formulate a textbook style review, the second stage is to acquire the ability to formulate a critical, reflective review and the final stage is to acquire the ability to conduct an informed and reflective analysis of the new in light of previous work. Such a staged learning process might work in a discipline based on a given body of literature, but it becomes rather inefficient when a study deals with an emerging field where one of the key tasks is to define the relevant literature (see the section Relevant literature – your canon).[2] The problem formulation process in an interdisciplinary context is generally an organic process, entailing not only the identification of relevant literature but also the identification of perspectives and aims as well as the collection of evidence, and all of these processes are intertwined. In such a context, it is counterproductive and thus neither necessary nor desirable to learn how to summarize a field in a textbook style.

The terms "informed" and "reflective" in Question 9 are ambiguous and need to be interpreted. To do so, you must be aware of and learn how to navigate the marked style differences among disciplines. To complicate things, the term "analysis" is used in at least three different ways, which brings an additional challenge to interdisciplinary efforts. Before continuing, I therefore comment upon the different meanings of "analysis".

Defining "analysis"

Analyse = to gather information

The term "analysis" may refer to the technique used to collect or gather information. For example, it is possible to analyse the pH, redox potential, carbon content of a soil or water sample. Various types of "analytical instruments" are used to assist in the data collection procedure and the result of a specific analysis is generally numbers (if the study is quantitative) or a description of features of the object of study that the used instrument is able to detect (if the study is qualitative). The first excerpt below is from a study in which the authors have used three types of instruments to study (analyse) a sample, i.e. to determine its chemical composition. The "products are ... analysed by gas chromatography. ..." The second excerpt is from a study in which the authors have studied drainage water from an experimental site. The sentence "Routine analyses on the water draining the sandbox were performed weekly on a composite sample" tells the reader that the chemical composition of the water was determined with respect to the factors normally determined for field water samples:

Example 24
Analyse = to gather information

The fractionated products from a scaled-up reaction were extracted, acetylated, and **analysed*** by gas chromatography-electron impact mass spectrometry. The results showed that most of the products (II to VI) were uncleaved, chlorinated derivatives of model 1 (Fig.; Table 1). However, two of the minor products, VII and VIII, resulted from Ca-aryl cleavage of model I. HPLC and gas chromatography **analyses*** showed that these products cochromatographed with standards of 1-chloro-4-ethoxy-3methoxybenzene   and   1,2-dichloro-4-ethoxy-5methoxyben-zene, respectively.

Patricia Ortiz-Bermudez 2003, p. 5015

Soil organic matter content was determined by combustion at 550°C. Routine **analyses*** on the water draining the sandbox were performed weekly on a composite sample. Anions were determined by ion chromatography.

Shelley J. Kauffmann et al. 2003, p. 25

*bold emphasis inserted by the author.

Analyse = to process data

The term "analyse" is also used in a second sense. The procedure that follows the collection of data encompasses processing and interpretation of the data such that the results are made clear; this is often called analysing the data. This is especially common when applying statistical methods and this second step is often illustrated by a visual representation in the form of graph or table. Below is an example from a paper published in a renowned international journal in which the authors explain that they "analyse detailed series of chloride, a natural tracer, in both rainfall and runoff from headwater catchments in Plynlimon, Wales." This means that the authors have processed data obtained from a combination of hydrological observations and chemical analyses (sic!):

Example 25
Analyse = to process data

The time it takes for rainfall to travel through a catchment and reach the stream is a fundamental hydraulic parameter that controls the retention of soluble contaminants and thus the downstream

Continued

consequences of pollution episodes.[1,2] Catchments with short flushing times will deliver brief, intense contaminant pulses to downstream waters, whereas catchments with longer flushing times will deliver less intense but more sustained contaminant fluxes. Here we **analyse*** detailed time series of chloride, a natural tracer, in both rainfall and runoff from headwater catchments at Plynlimon, Wales.

James W. Kirchner et al. 2000, p. 524

*bold emphasis inserted by the author.

Analyse = to evaluate, interpret and discuss in light of previous research

A crucial part of an empirically based study is the evaluation, interpretation and discussion of the results in relation to previous research. This stage is usually what researchers refer to when the word "analyse" is used in the aim and, to a large extent, the analysis is what distinguishes a consultant report from a scholarly study. Below are two examples in which the word is used with the above meaning.

Example 26
Analyse = to evaluate, interpret and discuss in light of previous research

The aim of the study is to **analyse*** why and under what conditions co-operation between Israel and Palestinians and between Israel and Jordan has taken place and how it has functioned in the water sector. The study focuses on the water aspects of the respective peace negotiations as well as the implementation of what has been agreed upon. It therefore moves beyond the existing material which states that transboundary water co-operation does occur – material which is ample in a quantitative sense – by exploring why co-operation has occurred in the Jordan River Basin.

Anders Jägerskog 2003, p. 14

The most obvious difference between the results of the laboratory and field study was the transformations related to NO_3^- (Figs 2 and 7). It appears that sieving of soil disrupts processes related to the NO_3^- assimilation and subsequent use of NO_3^- in biosynthesis. For instance, the process of NO_3^- reduction to NH_4^+ is much more obvious in the field study (see unfumigated treatments in Figs 3a and 8a) than in the laboratory. Detailed **analysis*** of these processes in the context of N transformations will be the subject of a subsequent publication.

Christoph Müller et al. 2003, p. 663

*bold emphasis inserted by the author.

Clarifying the own, the new

As an academic writer, you must be able to clarify your own contribution, which seems to be one of the most difficult things to learn for most new-comers. It appears as if beginners in general assume that the reader intui-tively will understand when insights and conclusions are the outcome of the study in question rather than being the thoughts and insights of someone else. In the example below, Jane Lister discusses five mythical assumptions regarding sustainable forest management (SFM). We can guess that these five "mythical assumptions" have been identified by Jane, but it is not explicitly stated, so it is also possible that these five assumptions actually have been identified by someone else. If it is the former case, Jane risks not getting credit for the work. If it is the latter case, she risks being accused of plagiarism.

Example 27
Implicit presentation of findings

There is a "stickiness" or path dependency towards the traditional timber management paradigm that is anchored in mistaken ecological, social and economic assumptions and re-enforcing, rigid market and state-based institutions and organizations (Kant 2003). As a result, assumptions inherent in the traditional timber management paradigm, while generally accepted as invalid under the SFM paradigm, neverthe-less continue to misdirect forest management decisions.

Five of these "mythical" assumptions and the accompanying SFM reality are listed in Table 1 and discussed in the remaining sections of the paper. A hope is that by explicitly revealing these assumptions, not only will a set of SFM principles become clear, but also the accompany-ing challenge and necessity for institutional change will be evident.

Jane Lister 2007, p. 243

It is accepted in several traditions to present new thoughts, insights and conclusions implicitly. The example above is from such a tradition, and readers who are familiar with the field understand that the five mythical assumptions are the outcome of her work. When speaking to an interdis-ciplinary audience, there are at least two reasons to spell out the new ideas explicitly rather than implicitly, or at least as far as possible, to make it difficult for the reader to misunderstand what new insights a study has rendered. First, a study becomes more reliable if the reader can distinguish clearly between insights and conclusions that are the outcome of the study and those that are borrowed from someone else. If everything in a study is someone else's, it is but a comprehensive review, and to write a comprehensive review in an implicit manner, which might lead the reader to believe that it is a critical analysis presenting new insights, is nothing but plagiarism. Second, if a critical analysis is misread as a comprehensive review, the author will not gain the recognition s/he deserves for the original insights and conclusions that emerge in the study.

Relevant literature – your canon

To conduct an informed and reflective analysis in light of relevant literature, you must know how to draw the lines between relevant, less relevant and irrelevant literature. Within a clearly confined discipline or research field, it is rather clear what type of literature is considered basic and relevant, which journals you are expected to stay updated with and what books you are expected to have read. One of the major challenges when working in areas that cut across traditional academic borders is that no such common literature exists and, as a result, it becomes rather difficult to define "relevant literature" in a study that extends beyond traditional academic borders.

What to consider as relevant literature is in part defined by the aim of the study and how one has formulated the research problem, and in part a question of the audience. I will get back to the audience further on in this chapter in the section on taken-for-granted knowledge. As discussed in Chapter 7, the problem formulation process is most often iterative and the researcher moves between the general and the specific, between theory and method, between the abstract and the concrete. By iterating between these levels, it becomes possible to hone the aim, which is necessary to make the study feasible but still interesting from a general perspective. In the process of honing the aim, the literature plays an indisputable role, as illustrated by the citation below:

> I found that a research question gets continually pared down as you progress, even after you have started your proposal and research design. A question with a narrow focus enables a more directed research design, and therefore provides more rewarding and informing results. I also gained valuable insight into the role the literature review plays in problem formulation and research design. A review of current and fundamental literature forms the basis of the theoretical background for the study. The purpose of the literature review is to not only base yourself within relevant literature, but also to gauge it critically and decide

> how (or if) it is helpful to your study. My previous impression of citing
> literature and methods has now changed from interpreting previous
> studies conducted in my field as the *status quo* or something to strive
> for to seeing them as perspectives and deductions I can use to frame
> my work and my conclusions as I see fit.
>
> *Kimberly Jones 2007, with permission from the author*

Kim, who also is a former graduate student from UBC, emphasizes the difference between reading to acquire a broader scholarly base and actively reading to assess whether or not the literature is useful for the study in question. She also emphasizes the difference between reading to become familiar with a certain body of literature and viewing the literature as examples of perspectives and conclusions, which may or may not be used to frame her study. By critically examining the literature in light of the aim of the study, you will thus be able to distinguish literature that provides general knowledge within a field from literature, which may be useful to frame a study.

The inexperienced student easily falls into either of the following two traps: reading everything under the impression that "everything is relevant" or reading nothing under the impression that "no one has written about this before". If you find yourself with a reading list that is too long, you have probably defined the aim too widely and put too little emphasis on the specific, the concrete and the method. In contrast, if you do not find any relevant literature you have probably put too much focus on the specific and too little emphasis on the general, the abstract, and the theory.

I suggest that you return to the three papers you picked in the exercise in Chapter 6. This time use two markers of different colour to, on the one hand identify sections that deal with the general issue at hand and on the other hand identify the sections dealing with the specific issue. Following the procedure outlined before, meet with a group of students who have done the same exercise (with their papers of choice) and discuss your findings. This will help you get a firmer grip of what it means to define and demarcate a study on a general as well as a specific level.

It is crucial to understand that an academic study cannot be relevant only to the case in focus – it must also, in one way or another, shed light on something from a general perspective. The question is what this "something" is. Learning how to design a study that enables a deeper understanding of a general issue at the same time as it sheds light upon the case in question is a central part of good academic scholarship training. The trick is to learn to define and demarcate the study on both specific and general levels. You may start with either, but it is advisable to iterate between the two until you have reached a design that enables a deeper general understanding as well as a deeper understanding of the specific case.

As mentioned above, finding oneself with nothing to read is usually the result of defining "relevant literature" as literature that deals with the specific case in question. To illustrate, I use a student who is to investigate whether or not the concentration of metals in a specific river has changed, let's say in Motala Ström in Sweden. An inexperienced student has a tendency to look for literature that deals with heavy metals in Motala Ström. The student might be able to find some consultant reports but probably little if anything in the international literature. This is because a study that focuses on Motala Ström and refers only to literature on the very same river might be a good consultant report but, since such a study does not elucidate anything from a general perspective, it does not meet the standards of an academic study.

To transform the study from an interesting consultant report which is useful for the specific case to an acceptable academic work that also elucidates a general issue, the student could focus on changes of heavy metals in rivers over a specific period of time, say from the 1950s to the present. Data from Motala Ström could then be analysed in light of previous literature dealing with changes in heavy metal concentrations in rivers around the world during the given time period. The specific question is: How and why have the heavy metal concentrations varied in Motala Ström? And the general question: Why do heavy metal concentrations vary in surface water? The general question may be narrowed down by focusing on rivers with environmental characteristics similar to those of Motala Ström, and then the general question would be: Why do heavy metal concentrations vary in surface water with xx characteristics?

I illustrate with another example: a student who wants to study the implementation of the Environmental Bill in Sweden. The tendency here may be to focus only on literature dealing with the implementation of the Swedish Environmental Bill. Since the bill is of a rather recent date and unique to Sweden, the literature defined as relevant becomes very limited. The specific level is "Implementation of the Swedish Environmental Bill". The general level could be defined as "Implementation of Environmental Bills" or "Implementation of Policy Bills in Sweden"; either way opens things up to make a wider literature available.

It is often useful to initiate dialogues that focus on distinguishing the specific level from the general level and, above all, to identify from what perspective the chosen subject is interesting on a general level and then identify relevant literature from the point of departure of the chosen perspective. Such an approach strengthens the ability to use the chosen perspective to search for relevant literature and to use the literature to sharpen the aim of the study.

Common knowledge

All academic authors presuppose that readers share a common knowledge base with them, and it is considered rather obscure to provide references when referring to common knowledge. This means that you are expected to be literate in your field. The meaning of "common knowledge" does, however, vary among discipline, which poses yet another challenge for the interdisciplinary researcher. Generally speaking, you will refer only to common knowledge literature when a widely accepted theory is challenged. Some disciplines demand that you read original sources but, in most cases, key sources or classical texts are known only through secondary sources. For example, with a BSc in biology, you are expected to be familiar with Charles Darwin's *On the Origin of Species*, someone with a BSc in physics is expected to be familiar with Einstein's theory on light, energy and mass ($e = mc^2$), and students of political science are expected to be familiar with Plato's *The State*, but in most cases the undergraduate students are not expected to actually read the original work; it is considered sufficient to recognize the title, the theory and the author and tie these to major concepts and ideas of the work. It is, for example, very rare to find references in contemporary research in evolutionary studies, atomic physics or political science to Darwin's, Einstein's or Plato's original work, although studies in each of these fields rest upon and are expected to be familiar with the general ideas of the theories outlined by these scholars. In addition to classical texts such as those mentioned above, every field rests upon a body of literature, often implicitly understood as the common knowledge of the field – the Canon. The challenge for any interdisciplinary scholar is that there is no given canon, especially if you work in an emerging field. Part of your work will therefore be to establish your canon. Return once more to your three papers: what do these authors assume that everyone knows? What do the authors of the papers picked by your student fellows assume as common knowledge?

Original research

An ongoing academic analytical discussion is at the heart of all research traditions, and the discussion is carried out both in oral and in written forums. The oral discussion is generally carried out in seminars, at workshops and at conferences. In some traditions, the written discussion is mainly handled through books, whereas other traditions carry out the discussion through peer-reviewed journal articles. Most disciplines require that you refer to the written discussion, although you on occasion may be allowed to refer to oral sources.[3]

In traditions where the discussion is carried out through books, one book assumes a certain perspective, criticizing and discussing other perspectives and conclusions while another book, assuming another perspective, discusses and criticizes the first book. Hence, knowledge presented in books is continuously debated and questioned and the ability to participate actively in such discussions is a crucial part of the undergraduate training. In order to become familiar with current academic disputes, you must read a number of books in each field to become acquainted with the subject. In such disciplines, there are no textbooks that summarize the field. "Common knowledge" means that you are expected to have read the original literature and you are usually expected to spell out the literature that is used as a basis for the discussion, as illustrated in the example below:

Example 28
Explicit reference to the canon

Introduction

/.../ This ideology is called neo-liberalism or globalism; due to the international character of present-day economy. One of the major critiques of neo-liberalism is that it does not differentiate between politics and economy. Politics is foremost seen as a tool to create advantages for the business sector, which strives to maximize its benefits.[1] The neo-liberalist school argues that free trade is the solution to sustainable development, as competition will solve social problems and environmental degradation. In opposition to that it can be argued that the world isn't that simple. The critics of globalism point out that in developing countries today, with an economy based on export, the prevailing ideology has resulted in social and environmental degradation.[2] This situation creates a struggle between human and environmental rights and the financial interest of business.[3] /.../

The liberalisation and deregulation of the Indian market began in the late eighties when the country's economy opened for international competition.[4] /.../ The fact that the environmental legislation has developed considerably during the past 25 years has had no or little affect on the industrial pollution.[5] The state has failed to handle pollution which may be due to the deregulation and in the wish to increase export. /.../ As a result the state has not been able to accomplish its

Continued

promised goals in various policy areas.[6] /.../ Approximately 70 percent of the available water in India is polluted.[7] There is a need to solve the challenge of developing an exporting industry while simultaneously protect the environment and the livelihood of the people. In this study I discuss the effects globalism on the local context in Tirupur, as an attempt to see the situation in Tirupur in a broader perspective.

While the state and the market have failed to handle India's pollution problems, the management of water resources through Common Effluent Treatment Plants (CETP) may be an alternative. Throughout India, and in the textile industry town Tirupur, a national approach of building CETPs have been introduced as an alternative trying to handle severe pollution of water.[8] The set up of CETPs have been seen as an initiative to control industrial pollution for clusters of SSI.[9] When constructing and operating a CETP, various industries have to collaborate. Blomqvist identifies CETP in Tirupur as a local entity of collective action.[10]

---

[1]   Beck, p. 23
[2]   ibid., p. 153 f
[3]   Shiva, p. 33
[4]   Tropp 1998, p 80
[5]   Blomqvist, p 129
[8]   Tropp 1998, p 16
[9]   Sapru, p. 156
[10] Parikh, Radhakrishna, p. 142 f
[11] Blomqvist, p. 44

Johanna Eriksson 2003, pp. 6–7

In the example above, the master's student places her study in a post-modern, neo-liberal critical framework by referring to the authors Ulrich Beck and Vananda Shiva. Beck is known for his analysis and critique of what he calls the modern risk society and Shiva is known for her critique of globalism and free trade as threats to sustainable development, particularly in developing countries. The paper would be deemed less rigorous if references instead had been made, for example, to a couple of master's theses taking their point of departure in the theories elaborated by Beck and Shiva, or if references were made to unpublished work by unknown authors. In these types of texts, referring to theories elaborated by internationally renowned scholars is a necessary distinguishing mark of a credible research text. After introducing the reader to the general framework and overarching theories, the author generally continues to more specific issues, here the relation between sustainable development and environmental degradation in India, by referring to books and book chapters dealing with these issues, in the excerpt exemplified by a book by Parik and Radakrishna and a chapter by Sapru. Finally, the study moves from the Indian context to site-specific information, i.e. information about Tirupur as given by Blomqvist. Hence, the study is framed and rooted in previous research all the way from the general theoretical framework to the specific site and situation of the actual study.

Textbooks

In contrast to the traditions described above, other disciplines summarize the theoretical framework in textbooks. The differences among textbooks in such fields are mainly a question of which parts are emphasized rather than the theoretical framework itself. Students at the undergraduate level usually read only one textbook in each subject and are rarely introduced to disputes of current interest before graduate level. Current disputes are carried out in articles published in the peer-reviewed literature. The canon in such traditions does consequently consist of a few basic text books and a large number of journal articles. In such disciplines, it is considered obscure to refer to textbooks in academic papers unless a generally accepted theory is challenged. Hence, references to so-called textbook knowledge in such disciplines are generally omitted in written documents.

Example 29
Implicit reference to the canon

Although anthropogenic sources are currently considered the primary origin of organohalogens in the environment, studies conducted in the last several years have indicated a major natural source of these products.[1,2] One feasible source of natural organochlorines in soil systems is the chlorination of plant material during weathering. Unexpectedly high, temporally variable concentrations of organohalogens in decaying forest leaf litter and in soils remote from anthropogenic sources indicate that both halogenations and dehalogenation are related to natural organic matter decomposition.[3] Fresh plant matter has been analyzed by Nkusi and Muller[4] and Flodin et al.,[5] with quite disparate results. Nkusi and Muller[4] determined that up to 75% of halogens in one terrestrial plant species (*Euphorbia charasias*) were present as organohalogens, while Flodin et al. found a maximum of 10% organohalogens

Continued

in fresh sprucewood (*Picea abies*) and less in fresh sphagnum moss and birch leaves. More recently, *in-situ* X-ray absorption spectroscopy (XAS) studies have indicated that the concentration of chloride (Cl⁻), the dominant form of chlorine in fresh plant material of several plant species, appears to decrease while organochlorine concentration increases as the plant material weathers naturally in the environment.[6] However, organic molecule chlorination reactions in weathering plant materials remain poorly understood.

---

[1]   Asplund, G., Grimvall, A., Pettersson, C. *Sci. Total Environ.* 1989, 81/82, 239–248.

[2]   Gron, C. In *Naturally-Produced Organohalogens*; Grimvall, A., DeLeer, E. W. B., eds., Kluwer Academic Publishers: Dordrecht, The Netherlands, 1995; pp. 49–63.

[3]   Oberg, G., Nordlund, E., Berg, B. *Can. J. For. Res.* 1996, **26**, 1040–1048.

[4]   Nkusi, G.; Muller, G. In *Naturally-Produced Organohalogens*; Grimvall, A., de Leer, E. W. B., Eds.; Kluwer Academic Publishers: Dordrecht, The Netherlands, 1995; pp. 261–268.

[5]   Flodin, C.; Johansson, E.; Boren, H.; Grimvall, A.; Dahlman, O.; Morck, R. *Environ. Sci. Technol.* 1997, **31**, 2464–2468.

[6]   Myneni, S. C. B. *Science* 2002, **295**, 1039–1041.

Rachel Reina et al. 2004, p. 783

In the example above, the reader is informed that the study deals with processes that take place during the degradation of litter (weathering of plant debris). In line with the tradition, no reference is given to an overarching framework. The informed reader is expected to know that general information on the subject may be found in textbooks on biogeochemistry and soil science and more specific but still overarching information in textbooks on degradation processes. Instead, the framing of the paper concentrates on a rather specific level, in this case chlorine cycling, focusing on the news value of the presented work. With a silent voice, the authors inform the reader that the field is new and assures that they are well acquainted with the little there is to know; this knowledge is emphasized through phrases such as "have indicated /.../", "One feasible source /.../" "quite disparate results /.../." The paper would be seen as less rigorous, or even unreliable, if it was found out that a doctoral thesis not referred to in the paper had been devoted to the chlorination of organic molecules in weathering plant material or that a paper had been published on the subject in a renowned journal.

> **Example 30**
> **Reference to a textbook**
>
> It is well documented that chloride participates in a complex biogeo-chemical cycle (Neidleman and Geigert 1986; Asplund and Grimvall 1991; Grimvall and de Leer 1995; Gribble 1996; Winterton 2000; van Pee 2001; Laturnus *et al.* 2002; Öberg 2003). Still, it is widely believed that all chlorinated organic compounds are xenobiotic, that chlorine does not participate in biological processes, and that it is present in the environment mainly as chloride ions (Cl_{inorg}) (e.g. Schlesinger 1997).
>
> *Gunilla Öberg et al. 2005, p. 173*

The example above is from one of my own papers and we cite a textbook (Schlesinger) because the study challenges a generally accepted theory. We clarify for the reader that a prevailing theory is challenged since evidence is presented in support of a new theory by using a terminology that is easily recognized by our audience; we first state that "It is well documented that ...," supporting this statement with references to a number of recent studies and then contrasting these with "Still, it is widely believed...," referring to the prevailing theory as presented in William Schlesinger's widely used textbook *Biogeochemistry*.

The style of recognized scholars

The prose style used by recognized authors, especially in the humanities and qualitative social sciences, usually entails small, tacit hints that are clear for those familiar with the tradition and which show that the author knows his/her way around. These hints are to some extent provided through the choice of words, but most of the credibility of the text comes from the way the author refers to previous sources. The rigour of the method is most often embedded in the text and there is generally no shortcut to assessing the credibility: you must be familiar with the methods, you must be familiar with the field and you must read the whole text. I use the book *Re-thinking Science* by Helga Nowotny, Peter Scott and Michael Gibbons as an example. The three authors are renowned senior scholars in the STS-field[4] and their book clearly targets that field. On the back cover of the book, Wolf Lepenies, Rektor[5] at the Institute for Advanced Study in Berlin, is quoted:

> What you always wanted to know about the "knowledge society" is laid bare by Nowotny, Scott and Gibbons in *Re-thinking of Science*, the sequel to their much acclaimed book *The New Production of Knowledge* (1994). This is a splendid book, full of empirical insight and intellectual vision. *Re-thinking Science* is reliable and robust at the same time.
>
> *Wolf Lepenies 2001*

Those unfamiliar with the field are likely puzzled by the statement that this book is "full of empirical insight", as it is rather unclear for the newcomer what exactly the authors have used as their empirical basis. The entire book seems to be based on "opinions" rather than empirical evidence, and numerous claims are made that seem to be loosely grounded, if grounded at all. For example, on page 67 the authors write:

> Nevertheless, as in any co-evolutionary process, some remarkable convergences can be observed. On the side of science it is above all its remarkable efficacy to produce novelty (given sufficient resources, personnel, the right environment and its other important "material" inputs like concepts, techniques, experimental setups) that now receives the greatest emphasis. On the side of society this efficacy is now being fully recognized; there is an explicit willingness to exploit this efficacy to pursue various socio-economic goals. The linchpin of these co-evolutionary strands, the connecting node that binds them together, is innovation.
>
> *Novotny et al. 2001*

For the outsider it may appear as if this text totally lacks anchoring. For example, what is the basis of the claims that "some remarkable convergences can be observed," that "science has a remarkable efficacy to produce novelty" or that "society now has recognized this efficacy?" What are these statements based on? For those familiar with the field, these questions are trivial as they deal with "common knowledge" issues. The reader is expected to be well acquainted with the common knowledge literature in the field, which includes previous works of the authors. The book is *implicitly* anchored in a specific tradition and it speaks to those who are familiar with this tradition. Readers who are unfamiliar with the authors' previous work along with work of others, for example Bruno Latour, Ulrich Beck, Niklas Luhmann, and Karin Knorr Cetina, will not be able to understand why this is lauded as an exceptional and "splendid" piece of work.

People who are not scholars in the humanities are often unable to decipher the rigour in such a text and the reason is that the signs that make these texts credible are often so tacit that they are sometimes not even understood by undergraduate students, only by advanced scholars. This makes it possible for the inner circle to identify and denounce studies conducted by non-scholars, but the way is also paved for the widespread misunderstanding that studies in social science and the humanities, in contrast to other academic areas, do not demand good craftsmanship. It is simply not possible for the outsider to distinguish between scholarly work like *Re-thinking Science* from novels and journalistic texts. To the insider, the signals of rigour are clearly spelled out on every page of the book; to the outsider it is a text filled with opinions and no rigour. For renowned scholars such as Novotny and co-workers, this is hardly a problem. In contrast, for you as a newcomer, especially if you work in an emerging field or conduct work in a cross-boundary field, it is crucial that your text clearly signals rigour to a broader group of scholars and not only to a specific discipline.

The tradition to not explain clearly, and in simple terms, what you have done and what the arguments are based upon, makes it difficult or even impossible for an outsider to appreciate the level of the work, which thus often is underestimated. The implicit reference system works fine when the goal is to reach a well-defined audience, as in the case with Nowotny and co-authors, who clearly target the STS audience. However, when the audience is less clearly defined, as in most cases for interdisciplinary scholars, the implicit system is no longer efficient and a more explicit style is not only advisable but necessary.

Passive and active voice

There are different ways to clarify the roots of a piece of information, and different disciplines favour different styles.[6] Hence, while a student in a disciplinary setting only needs to figure out "how it's done", the interdisciplinary student needs to make an active choice, since there's no given way to "do things right".

"Voice", or, expressed slightly differently, the way the author chooses to convey the message to the audience, is one distinguishing characteristic. In the excerpt from Torgny Lindgren's *Hash* in Chapter 9, Lindgren has chosen to use the voice of a (fictitious) narrator to describe (fictitious) events, which are reported in the (existing) local newspaper of the (existing) small village where the (in central parts fictitious) story is played out. Fictitious narrators are to my knowledge not used in academic writings, but the voice of the author can be explicit or implicit, active or passive. To help you choose a form that suits you and that is accepted by those audiences you target, I suggest you return to the three papers you picked in Chapter 6. How would you describe the voice of your authors? What are the signs that you use as a basis for your description? What is the preferred style in your field of research or in the audience/s you are targeting? What are the distinguishing marks of this style?

One of the graduate students at IRES, Jane Lister, generously shared a piece of her text before and after discussing active and passive voice with her supervisor:

> By the way, I can give you lots of good examples of the huge difference changing from passive to active voice. Gave Peter my edited chp yesterday and he raved. WAY better now that I've taken ownership through voice, e.g.:

Old "Passive Voice" Version

> .... the classification of certification as a non-state market driven mechanism is then reviewed and the dual identity of certification as a global governance mechanism as well as an alternative "hard law" policy instrument at the domestic level is explained. The interaction of certification private authority with public authority within both the domestic and international arenas is discussed before turning to the next section which specifically examines the co-regulatory dynamic of certification as a complementary private policy instrument within a hybridized policy mix of traditional command-control regulation, delegated modes of self-regulation and emerging forms of regulated self-regulation....

New "Active Voice" Version

> .... I then review the classification of certification as a non-state market driven governance system and assess how certification private authority and traditional public authority interact within both the domestic and international political arenas. The next section analyzes certification as a co-regulatory policy instrument. Specifically, I evaluate how certification standards overlap with command-control regulation, delegated modes of self-regulation and emerging forms of regulated self-regulation....

Jane Lister, May 2008, e-mail communication.
Reprinted with permission from author

In the example below, the authors sometimes use an active voice and sometimes a passive voice. For example, in the sentence "Cox *et al.* (2000) obtained ...," "Cox and co-workers" are clearly active agents and sort of "step into" the text. In contrast, when referring to earlier studies, such as in the sentence "The warming signals in Levis *et al.* (2000), /.../ were small...," "Levis and co-workers" are passive agents hidden behind the text (as the text refers to warming signals observed in the study conducted by Levis and co-workers, and not to warming signals in Levis). Further down in the text, the authors themselves are active agents, talking about themselves as "we" (3rd line from the bottom of the paragraph).

Example 31
Mixing passive and active voice

In a simulation that did not include the radiative effects of increasing CO_2 concentrations, Cox et al. (2000) obtained a slight warming over land by the year 2100 due to CO_2-induced changes in stomatal conductance and vegetation distribution. In their coupled climate-carbon model simulation, Notaro et al. (2005) obtained a global mean warming of about 0.1 K during the historical period (pre-industrial to present day) due to the physiological effects of increasing atmospheric CO_2. The warming signals in Levis et al. (2000), Cox et al. (2000) and Notaro et al. (2005) were small because the simulations were run for only a few decades or a century. However, dynamic changes in vegetation distribution could take many centuries. Using a coupled climate-carbon cycle model that simulates vegetation dynamics (Bala et al. 2005; Govindasamy et al. 2005), we compare two multicentury simulations in this study: a "Control" simulation with no emissions, and a "Physiol-noGHG" simulation with prescribed emissions.

G. Bala et al. 2006, p. 621

In contrast, the authors of the following example use the passive voice throughout the text, both when referring to the work of other authors and when speaking about their own work.

Example 32
Exclusively passive voice

The watershed is a natural unit for water research (Clark and others 2005). Watersheds around the world are facing serious threats to their water quality and aquatic ecosystems. Moreover, watershed management, which includes water resource utilization control, water pollution control, and economic growth policies, is an effective means of dealing with these issues at the watershed scale (Heathcote 1998). Owing to the complexity of watersheds, uncertainty is one of the key factors influencing watershed management. Many methods have been employed to deal with the uncertainties in watershed systems (Guo and others 2001; Do' rner and others 2001; Sohrabi and others 2003; Chau 2004; Hamed and El-Beshry 2004; Ogata 2004; Zou and others 2004; Hantush and Kalin 2005; Muleta and Nicklow 2005; Zheng and Keller 2006).

Yong Liu et al. 2007, p. 678

Voice is clearly a question of taste, and different disciplines favour different styles, often strongly articulated. For example, Robert A. Day and Barbara Gestel clearly express a preference for the more active voice, simply claiming that a passive approach is "wrong":

> Some authors get into the habit of putting all citations at the end of sentences. This is wrong. The reference should be placed at that point in the sentence to which it applies. Michaelson (1990) gave this example:
>
> We have examined a digital method of spread-spectrum modulation for multiple-access satellite communication and for digital mobile radiotelephony. (1,2)
>
> Note how much clearer the citations become when the sentence is recast as follows:
>
> We have examined a digital method of spread-spectrum modulation for use with Smith's development of multiple-access communication (1) and with Brown's technique of digital mobile radiotelephony. (2)
>
> *Day and Gestel 2006, p. 80*

There is a spectrum from the passive to the active voice. In the example provided by Day and Gestel, the authors use "we" and are thus active agents in their text. However, the origin of their information is de-personified. The researchers who have conducted the studies to which the authors refer are not explicitly mentioned. It is implicitly understood that the information originates from the sources 1 and 2 but it is not clarified which information originates from which source. In the rewrite, the information is clearly tied to Smith and Brown, respectively, who in this style are made active agents in the text. Some disciplines strongly favour variants of a passive voice, and many do not allow the use of "I" or "we". There is no clear-cut line between the humanities and the natural sciences, although I have run into some natural scientists who adhere to the view that this is a common trait of *all* natural sciences. Examples of the use of the active voice, the passive voice, and a mixture of the two appear in various disciplines in all fields. The choice of passive or active voice is, to my experience, a very common source of conflict. The easiest way to solve such conflicts is to discuss the matter explicitly with the point of departure in some pertinent examples.

Notes

1   More on style further below in this chapter.
2   More on identifying relevant literature further below in this chapter.

3 Oral references should not be confused with references made to people one has interviewed as part of the empirical data collection procedure. The former is a question of anchoring the analysis in ongoing or previous research; the latter is a question of presenting one's own original research results.

4 Science and Technology Studies focuses on knowledge production in Science and Technology and its aim is described by York University's website as "the purpose is to expand our understanding of science and technology by exploring their social, cultural, philosophical and material dimensions. To achieve that purpose, the programme draws upon the disciplines of the humanities and social sciences to offer courses treating specific scientific ideas, as well as courses addressing broader topics such as science and gender, science and religion, and technology and cultural values. Students are encouraged to draw connections across traditional boundaries as they seek an intellectual appreciation for the sciences and technology as powerful means for understanding, embodying and shaping the world and ourselves. Students will learn to analyse complex ideas about science and technology, and to discover how to trace the origins and implications of events and patterns of thought in the past and present." http://www.sts.yorku.ca/ (8 February 2008).

5 "Rektor" in, for example, German, Swedish and Norwegian academia, is the same as Vice-chancellor in the British system and President in North American institutions.

6 More about style in Janet Giltrow 2002.

11 Beauty is in the eye of the beholder

Surprisingly, the form and structure is one of the issues that often cause difficult conflicts in interdisciplinary educational and research projects. Even though all in academia agree that an academic text should be concisely structured, there are diverging ideas of where, in what way and how extensively you are expected to address various issues. The conflicts generally do not deal with *what* but rather *where* and *how*. The tenth and last question in the framework is designed to address this issue: Are the form and structure in line with agreed norms? In this chapter, I discuss form in general but focus on two issues that far too often cause discord in interdisciplinary settings: headings and references. The goal is to help you become aware of different ways of presenting your text and consciously choose a way that suits you and your audience. As a start, I suggest that you once again return to your three papers. The two following exercises are quite simple compared to the earlier ones in Chapters 6 to 10, but they are worthwhile as it will help you gain a broader understanding of different styles and requirements, and will help you waive critique from less informed colleagues:

1) Describe how the headings in your three papers are structured, meet with your student fellows and compare notes. For your thesis – pick a form that you like and that caters to the audience you wish to reach. Later, when you choose a peer-reviewed journal, check what style they propose and stick to it.

2) Describe how the references in your three papers are structured, meet with your student fellows and compare notes. For your thesis – pick a form that you like and that caters to the audience you wish to reach. When you are ready to publish in a peer-reviewed journal, check what style they propose and stick to it.

Interdisciplinary Environmental Studies: A Primer, 1st edition. © Gunilla Öberg.
Published 2011 by Blackwell Publishing Ltd.

It seems to me as if academia is immersed in disciplinary snobbism. I apologize for whining, and it is not my intention to insult anyone, but I have to admit that I can actually not count the times I have heard academics express horror over the quality of theses, articles, books and research papers from other departments or research fields. And I cannot count the times when it has become embarrassingly clear that the one who has spoken had only leafed through the piece in question and made a judgement after a quick glance. I share the view that it is possible to in part judge the quality of a manuscript after a quick glance – in your own field. It is unfortunately quite time consuming to evaluate the work of other disciplines. My experience is that many scholars underestimate the time and effort needed to learn how to appreciate and evaluate forms other than those of your own tradition. You must be prepared to meet scholars who are unable to appreciate the form and structure used by others. Thus it is all the more important that you be aware and accept that a credible text may be structured in various ways, and that you be able to explain in what way the form you have chosen is acceptable.

A successful thesis written in an interdisciplinary environment will inevitably be coloured by a number of disciplines. Your text will probably take a form that is not really familiar to any of the more senior scholars involved and is even less familiar to people outside the group. In order to gain recognition outside your environment, it is crucial that you make conscious choices not only regarding aim, theory, method, evidence and analytical framework, but also regarding the form and structure of your work – and to consequently and consistently follow the form and structure you have chosen. To be able to choose, you must be aware of the options and you must acquire sufficient skills to manage the style you choose.

The preferred format of a thesis in a specific discipline is to a large extent coloured by the form and structure of the documents that scholars produce to qualify themselves in the academic system; some traditions demand publishing of articles, while other traditions are based on the production of reports or books. Disputes on headings and reference management are common and if not possible to avoid, at least handle them by increasing your level of awareness and preparing yourself for a clarification.

Headings

Headings are an area in which there are numerous different and very, very peremptory opinions and it is yet another area where the disciplinary student (generally) only needs to figure out the "right way" while the interdisciplinary student has to make an active choice. The major dividing line seems to deal with to what extent you should use standardized headings. To illustrate some of the various ways to formulate headings, I describe some general characteristics of the two extremes: standardized first-order headings and thematic headings. The description will by necessity be a rough generalization, but I hope to bring increased insight to the fact that things can be done in various ways and still result in credible research texts.

Standardized or thematic headings?

Standardized headings are descriptive and focus on the type of information rather than the information *per se*: "Introduction", "Material and Method", "Results", "Discussion" and "Conclusions". In many disciplines, perhaps most pronounced within the experimental, laboratory based natural sciences and medicine, there is one and only one way to write first-order headings, often referred to as the IMRAD style. With small variations (i.e. "Method and material" instead of "Material and method") the headings follow the same pattern. Scholars in this tradition qualify mainly by publishing academic articles in peer-reviewed international journals, and doctoral theses do often have the form of compilation theses; the basis of these consists of a number of articles that either are published, have been accepted or submitted, or are to be submitted to a peer-reviewed international journal. Among these traditions, the academic journal format sometimes serves as the model for papers written at undergraduate level. The tradition of standardized first-order headings is mirrored in instructions for authors in many international journals, as in the example below from the author instructions of the international journal *Chemosphere*.

Example 33
Standardized headings

Article structure

Subdivision – numbered sections
Divide your article into clearly defined and numbered sections. Subsections should be numbered 1.1 (then 1.1.1, 1.1.2, …), 1.2, etc. (the abstract is not included in section numbering). Use this numbering also for internal cross-referencing: do not just refer to "the text". Any subsection may be given a brief heading. Each heading should appear on its own separate line.

Continued

Introduction
State the objectives of the work and provide an adequate background, avoiding a detailed literature survey or a summary of the results.

Material and methods
Provide sufficient detail to allow the work to be reproduced. Methods already published should be indicated by a reference: only relevant modifications should be described.

Theory/calculation
A Theory section should extend, not repeat, the background to the article already dealt with in the Introduction and lay the foundation for further work. In contrast, a Calculation section represents a practical development from a theoretical basis.

Results
Results should be clear and concise.

Discussion
This should explore the significance of the results of the work, not repeat them. A combined Results and Discussion section is often appropriate. Avoid extensive citations and discussion of published literature.

Conclusions
The main conclusions of the study may be presented in a short Conclusions section, which may stand alone or form a subsection of a Discussion or Results and Discussion section.

Appendices
If there is more than one appendix, they should be identified as A, B, etc. Formulae and equations in appendices should be given separate numbering: Eq. (A.1), Eq. (A.2), etc.; in a subsequent appendix, Eq. (B.1) and so on.

Chemosphere, Guide for authors, 2010

In contrast to standardized headings, thematic headings focus on the content rather than the type of content. A number of disciplines mix standardized headings with thematic headings. This is common, for example, among field-based natural sciences as well as among the social sciences. Scholars in these traditions qualify mainly by writing reports and books whereas articles appear to be of less importance for qualification, at least among some traditions such as geology and sociology. Doctoral theses are often monographs, although compilation theses do exist and the reasoning in most texts is more elaborated as compared to the stricter article format.

The less strict demands regarding headings are mirrored in many of the journals of these areas. For example, the journals *Studies in History, Philosophy of Science, Meteorology and Atmospheric Physics* and *Ocean Dynamics* do not mention anything about headings in the author instructions. Instead the instructions are on a rather overarching level, here illustrated by the author instructions from the journal *Environmental Science and Policy*.

Example 34
No given format for headings

The manuscript should be organized in the following order:

Title page with authors' contact information.

The corresponding author should be identified with an asterisk and footnote giving e-mail address and telephone number. Full postal addresses (but not e-mail or phone) must be given for all co-authors.

Abstract: This should be of a maximum of 200 words, clearly stating methodology and the original contribution claimed in the paper.

Keywords: A list of up to 5 keywords

Main body of text: Authors should consult a recent issue of the Journal for style if possible. The Editors reserve the right to adjust style to standards of uniformity.

Appendices

References: Please see section III for more information.

Environmental Science and Policy, Instructions for authors 2010

Where do I place the reflections?

A common point of disagreement is in what sections of the text you are expected to make reflections. Among those that favour standardized first-order headings, some demand all reflection and analysis be conducted in the discussion section. According to these traditions, the results section should be a presentation only of the results, with no references made to other results or studies and thus clearly separated from the discussion. Others demand that reflections are conducted throughout the text, not least in the results section. Below I illustrate with excerpts from author instructions. The first example is from the highly ranked international journal *Soil Biology and Biogeochemistry*, which clearly belongs to the group who demand that the results section presents nothing but the results. In contrast, the likewise highly ranked journal *Environmental Science & Technology* in the second example asks for a joint presentation of the results and the discussion.

Example 35
Strict separation to between results and discussion

Results
This need only report results of representative experiments illustrated by Tables and Figures. Use well-known statistical tests in preference to obscure ones. Consult a statistician or a statistics text for detailed advice.

Discussion
This section must not recapitulate results but should relate the authors' experiments to other work and give their conclusions, which may be given in a subsection headed Conclusions.

Soil Biology and Biochemistry, Author instructions 2010

Example 36
No separation between results and discussion

Results and discussion
Here you can discuss your findings, postulate explanations for data, elucidate models, and compare your results with those of others. Be complete but concise. Avoid irrelevant comparisons or contrasts, speculations unsupported by the new information presented in the paper, and verbose discussion. Results and discussion may be combined or separated. Do not include a conclusions section in research articles. Include major conclusions in the abstract and in the body of the Results and Discussion sections.

Environmental Science & Technology, Author instructions 2010

Where do I describe the context?

Among some disciplines, it is common to demand one chapter named "Background" and another named "Theory". The question of "to be or not to be?" for the theory chapter is often a source of disagreement in interdisciplinary projects. The ability to navigate in an interdisciplinary environment may be easier if you are aware of this potential dispute over the theory chapter.

References

As discussed in Chapter 9, everyone agrees that one of the things that characterizes a credible academic work is that the author clarifies from where different claims originate, gives clear accounts for the sources and clarifies when the line of argument is based on the author's conclusions and when it is based on someone else's work. Negligence with sources will doubtless lead to a reprimand, since sloppily handled references in most disciplines is seen as an incontrovertible sign of non-credibility. It is crucial that you are consistent and use the same system throughout the text, irrespective of style. Surprisingly, many scholars are unaware that there are several ways to refer to sources and organize a reference list, and even more surprisingly, people may actually quarrel about such technical issues. You can fairly easily avoid wasting energy on such uncalled-for discussions by increasing your awareness of the differences and preparing yourself for a clarification. Below, I give some examples of differing ways to present in-text citations and to organize reference lists.

In-text citations

A common trait among almost all disciplines is that in-text citations are placed in the running text, most often as soon as possible after a statement, data or any type of information which is referred to that does not originate from the investigation at hand. The use of in-text citations is one of the traits that separate an academic text from other sorts of texts and it signals the living character of research texts. The text should enable the reader to evaluate the writer's use and interpretation of the cited sources continuously.

There are a number of different ways to clarify where a source originates from and most journals give clear instructions about what referencing system they prefer. In the first example below, the author has chosen a common variant that is often called the Harvard system: the authors surname and year of publication is given, all in parenthesis. When two authors, both are given; more than two authors is generally signalled by presenting the first and sometimes also the second author, followed by the abbreviation *et al*. The details may vary, as if and when to use period or comma, if *et al.* is to be in italic or not and if page numbers are to be given, as in the example below. The latter is more common when referring to books as compared to articles.

Example 37
In-text citation according to the Harvard system

Initially, anaerobic digestion seems like a simple chain of events in which a sugar molecule is broken down by various microorganisms to eventually yield methane and carbon dioxide (Cordina *et al.* 1998). However, it is a much more complex microbiological process than this and it has great potential if it is understood properly and if the knowledge acquired is optimized (Hooper and Li 1996). /.../ Taking this one step further, the resulting digested rest products can then be used to fertilize the land on which the grains are cultivated (Ejlertsson, pers comm,.). /.../ Rather, the initial substrates must be converted to simpler forms such as soluble monomers (Gujer and Zehnder 1983). /.../ These products then become substrates for the fermentation processes that follow (Zehnder 1988).

Anna Lundén 2003, p. 4

The Oxford system is another common system, the system chosen for this book: footnotes are placed at the bottom of the page, at the end of a chapter, or at the end of the book, and references to the footnotes are given by superscripted numbers. Footnotes may be used as in-text citations or be informative, which is very common in the humanities and more common in the social sciences than in the natural sciences. The footnotes are very detailed in some traditions. For example, in theology and history, it is not uncommon to find footnotes covering over half of the pages. Such footnotes are meant not only to help the reader find the reference; they are also a place to comment upon the text. Below follows an example where the footnotes are informative; after that follows an example where the author uses the footnotes to comment upon the text.

Example 38
Reference information in footnotes

The lack of progress in the policy negotiations was based on fundamental differences of opinion, not only with respect to what needed to be done to address the challenges posed by climate change, but also

Continued

regarding the relation between science and policy. For example, according to one interview account, there was an unwillingness of one of the parties to paraphrase or draw in any way from the scientific results as it would be an interpretation that had not been done by the experts.[196] This is in contrast to the preference expressed by the Inuit Circumpolar Conference for a document that combined explanations of the impact of climate change based on the scientific assessment with clear policy recommendations.[197] Another disagreement was on the form that any policy recommendations should take, i.e. should they be in a stand-alone document.[198] The disagreements in principle and the public airing of them appears to have created a very tense atmosphere, as exemplified by a letter from the chair of the Senior Arctic Officials that pointed out that the public disclosure of the closed meeting between Senior Arctic Officials and the Permanent Participants were fundamentally at odds with the practices in the Arctic Council. The response from the Inuit Circumpolar Conference chair was that her comments had been limited to "defending the integrity of the process."[199] Negotiations continued in October in Iceland. On October 30, 2004, the *New York Times* revealed the findings from the ACIA scientific process in a front-page article with the headline "Big Arctic Peril Seen in Warming", apparently based on a leak of the overview document.[200] Just a few days before the ministerial meeting and after several meetings, the Senior Arctic Officials, most of them accompanied by the respective country's chief climate.

---

[196]   Interview 63

[197]   Watt-Cloutier, *et al*. "Responding to Global Climate Change: The View of the Inuit Circumpolar Conference on the Arctic Climate Impact Assessment."

[198]   Interview 61, 63, and 64.

[199]   Letter from Gunnar Pálsson, Chair of the Senior Arctic Officials to Senior Arctic Officials and Permanent Participants, October 1, 2004, and Letter from Shiela Watt-Cloutier, Chair Inuit Cicurmpolar Conference to Gunnar Pálsson, Chairman of Senior Arctic Officials to the Arctic Council, October 19, 2004; both as quoted in Watt-Cloutier, *et al*. "Responding to Global Climate Change: The View of the Inuit Circumpolar Conference on the Arctic Climate Impact Assessment," 65.

[200]   Andrew C. Revkin, "Big Arctic peril seen in warming," *New York Times* Page 1, Column 1, October 30, 2004).

Annika Nilsson 2007, p. 140

Example 39
Commenting on the text in the footnotes

A simple practical problem within a single discipline began the line of inquiry that led to this book. As a university teacher of writing I was charged with preparing students to write academic essays for their courses in all disciplines. Since academic assignments bear a loose relationship to the writing done by mature members of the disciplines, a serious investigation of writing within disciplines promised to turn up information useful to teaching undergraduates. The investigation from the first was interdisciplinary by necessity, but only in a superficial sense, in that the writing examined came from a variety of academic disciplines. The concepts and analytical tools, however, did not extend beyond the typical repertoire of the English department.[1]

---

[1]   What constitutes the repertoire of the English department is no easy thing to categorize, nowhere codified, and nowhere discussed with methodological clarity. Rather, on the literary side it is embodied in the corpus of literary scholarship and criticism and in the seminar practices of textual discussion. Primarily it consists of close textual readings and historical contexting. The textual readings are all framed by recognition of traditional literary devices, and have been intensified by new critical insistence on the text in itself. However, other modes of criticism have suggested the application of interpretive frameworks from other disciplines, such as linguistics, psychology, sociology, anthropology, and philosophy. Such imported frameworks are justified in two ways: either they represent fundamental truths so that they cannot help but influence texts, or the writer on some level was aware of such ideas and constructed parts of the text upon them. [...]

Charles Bazermann 1988, pp. 3–4

The need to manage references – a point of full agreement

During my many years of involvement in interdisciplinary projects of various kinds, I have never ever overheard any disputes over the management of references. Everyone agrees that each in-text citation must appear in the reference list and that all sources listed in the reference list must be cited somewhere in the text. It is also generally held that the information in the reference list should be sufficient for the reader to fairly easily find the source, at least with the help of a librarian. There are ongoing discussions on how to best cite web sources,[1] but the demarcation lines here do not appear to be rooted in the research disciplines. Hence, I will not elaborate upon the use of references any further, since it is not a source of interdisciplinary miscommunication.

Notes

1   See, for example, http://www.library.ubc.ca/home/evaluating/

12 Being interdisciplinary

When I started to write this book, I called it "The Beginners Guide to Interdisciplinary Studies".[1] But calling it a guidebook suggests that it is possible to guide the newcomer through the territory. The crux of interdisciplinary work is that you enter an unknown territory that requires a protean sensibility. Rather than a compilation of guidebook tips, this book can be imagined as a series of signposts indicating directions to help you – the reader – to fulfil the journey. It is as much, or more, about attitudes than techniques. The key message is that you need to carve out your own route by getting the dialogue going with those around you and rather than to see heterogeneity as a hindrance, you must learn how to use it as a way to strengthen your work. I therefore chose to instead call the book *a primer*, since the aim is to get you started and to help you get others started.[2]

Throughout the book, I have stressed the need to create common ground, to look for commonalities and to find fruitful ways to draw on differences. In the introduction I criticized the literature on interdisciplinary studies for focusing on obstacles rather than opportunities. The fact that this literature speaks so much of obstacles is not a coincidence: the studies of interdisciplinary studies show that it is easy to become overwhelmed, sidetracked or discouraged by the differences. The take-home message is thus not that you should ignore the differences, or pretend that they do not exist but rather that you need to find ways to convert them from challenges to opportunities. One way is to accept that differences exist while exploring commonalities, as outlined in Chapter 2. It is however not possible to explore commonalities without common ground. And to create common ground, one needs to create a respectful environment, which is very much a question of attitude. Unfortunately the academic structures seem to feed disciplinary arrogance. In this final chapter, I point to some key issues that can counteract disciplinary arrogance and help the creation of a creative and respectful environment, and thus the creation of common ground.

Interdisciplinary Environmental Studies: A Primer, 1st edition. © Gunilla Öberg.
Published 2011 by Blackwell Publishing Ltd.

Creating an open and respectful climate

Many set out to convince others that their approach (read interest) is the most relevant, and far too few have their mind set on understanding the relevance of the other perspectives. The question ought to be: "What can I learn?" But the question is more often: "What can I teach?" If an interdisciplinary group manages to find ways to encourage humble and respectful attitudes towards other disciplines, then joint progress will increase tremendously. The other side of the coin is that fostering a humble and respectful attitude can be mistaken for the encouragement of self-abasement. If the arrogant vindicator believes his or her perspective is most relevant and that they must teach the others some basics before the discussion can even start, a participant lacking in confidence, on the other hand, focuses too much on learning from others, which makes them unable to contribute.

A colleague once told me that she was shaken when she as a student moved from one school, where students commonly got "A"s, to another where "A"s were only given on rare occasions. At first, she felt like the "B"s she was given were insults, that she was a failure, but when engaging in dialogues with fellow students, she came to realize that the "B"s encouraged her to ask for feedback – it gave room for improvement. The school system far too often teaches that feedback is equal to negative criticism, which is equal to failure. That being good is to prove that the criticism is wrong and you were right in the first place. The ever so confident and immensely curious three-year-old who asks "why" with never ending curiosity grows into an arrogant (and fearful) seventeen-year-old who feels like a looser unless she wins every argument. It seems like we are commonly fostered to believe that asking "why" is a sign of weakness rather than strength. Your success as an interdisciplinarian will clearly benefit if you nurture your inner three-year-olds' curiosity and fearlessness. I don't think it's possible to over-emphasize the need to encourage humble and respectful attitudes towards other's knowledge, traditions and experiences – still keeping one's self-confidence. Not being aware of your inabilities is a major threat to interdisciplinary co-operation, but not being aware of your strengths is just as problematic. Identifying your weaknesses and your strengths is crucial for success. In the words of David Bohm, this is a question of learning to engage in dialogue:

> The derivations of words often help to suggest a deeper meaning. "Dialogue" comes from the Greek work *dialogos*. *Logos* means "the word", or in our case we would think of the "meaning of the word". And *dia* means "through" – it doesn't mean two. A dialogue can be among any number of people, not just two. Even one person can have a sense

of dialogue within himself, if the spirit of dialogue is present. The picture or image that this derivation suggests is of a stream of meaning flowing among and through us and between us. This will make possible a flow of meaning in the whole group, out of which may emerge some new understanding. It's something new, which may not have been in the starting point at all. It's something creative. And this shared meaning is the "glue" or "cement" that holds people and societies together.

Contrast this with the word "discussion", which has the same root as "percussion" and "concussion". It really means to break things up. It emphasises the idea of analysis, where there may be many points of view, and where everybody is presenting a different one – analyzing and breaking up. That obviously has its value, but it is limited, and it will not get us very far beyond our various points of view. Discussion is almost like a ping-pong game, where people are batting ideas back and forth and the object of the game is to win or to get points for yourself. Possibly you will take up someone else's idea to back up your own – you may agree with some and disagree with others – but the basic point is to win the game. That's very frequently the case in a discussion.

In a dialogue, however, nobody is trying to win. Everybody wins if anybody wins. There is a different sort of spirit to it. In a dialogue, there is no attempt to gain points, or to make your particular view prevail. Rather, whenever any mistake is discovered on the part of anybody, everybody gains. It's a situation called win-win, whereas the other game is win-lose – if I win, you lose. But a dialogue is something more of a common participation, in which we are not playing a game against each other, but *with* each other. In dialogue, everybody wins.

David Bohm 1996, pp. 6–7

Hierarchies that impair

Science vs "the other"

The creation of respectful attitudes is counteracted by academic "pecking orders".[3] As discussed in Chapter 2, there is a widespread, implicit but sometimes clearly explicitly stated hierarchy in which the exact sciences are judged as "more scientific" than those which are context bound. This hierarchy is visible in the funding system and poses a pertinent challenge when working across the humanities/social science–natural science/technology gap. The dividing line is also visible in writings. For example, the dividing line between qualitative and quantitative methods is discussed at length in social science and humanities literature but never mentioned in natural science and technology literature, even though both quantitative and qualitative studies clearly exist in all of these areas. Literature discussing qualitative social science research procedures uses a lot of space to explain strengths and weaknesses of qualitative methods, whereas literature dealing with qualitative natural science studies rarely or ever even acknowledges that they are qualitative studies and that such an approach may have strengths and weaknesses. The natural sciences are presented as self-evident and "normal" and therefore do not need explanation or defence, while the humanities/social sciences are defined in relation to natural sciences as "the other", i.e. not normal, and must therefore be defended and explained to gain authority. This implicit hierarchy often pushes people in the humanities (and social sciences) into a defensive position. Rather than being a given component of studies in the environmental and sustainability fields, it seems to be necessary to stress that the humanities are *also* needed, as if they were an add-on. I did for example receive comments on earlier versions of this book that indicated that readers felt that I marginalized the possibility of humanists doing work on interdisciplinarity. Some readers apparently got the impression that I included the natural sciences and excluded the humanities – even though I did not mention either. In my eyes, this book is a clarion call for the need to draw more on the humanities. The book draws heavily on theories, concepts, etc. from the humanities. Apparently, if this is not explicitly spelled out, there is a risk that this will not be appreciated. So, I revised the manuscript and it is now spelled out that the humanities are one of several necessary ingredients.

With this strong bias (i.e. when speaking about research or academic studies the natural sciences are implicitly included but humanities are excluded unless explicitly spelled out that they are included), it is not surprising that students in interdisciplinary contexts write lengthy justifications when using a qualitative approach with a humanities perspective and no justification whatsoever when conducting a study from a natural science perspective. It can be argued that all students should learn the craftsmanship and be able to hand over traditions. And it is a tradition in humanities (and some social sciences) that detailed explanations are to be given when using a qualitative approach. This is not the tradition in the natural sciences. On the other hand, by allowing, teaching, or even encouraging this discrepancy, we strengthen and reproduce a pattern that implies that natural science studies are the norm and first-grade research and thus better than qualitative studies within the social sciences and the humanities, which are implied as "the exception" and thus second-grade research. No doubt, the field of sustainability and environmental studies needs both qualitative and quantitative research and it needs the humanities as much as it needs the natural sciences. One of the reasons we need the humanities is because this is where the deep knowledge on meta-cognition and reflection is cultivated, which in itself is the root of another damning hierarchy.

Those with "real" academic training vs "the ignorant"

Those who are well-informed on overarching methodological issues often look down upon those who have no such training. Courses in the humanities and many of the social sciences usually emphasize the need to understand the research process at a general level, which in these environments equals "real academic training". Upon graduation, undergraduate students are expected to have acquired not only a rather deep understanding of the research process but also an ability to reflect upon how this has influenced their own knowledge-producing practices. Such expectations are rare or rudimentary within the natural sciences, medicine and engineering. Literature on the research process in such settings either deals with how to write a research paper from a rather technical point of view,[4] with specific procedural problems dealing with a certain technique,[5] or with the history of science in a descriptive rather than reflective manner.[6] Consequently, science and technology education rarely embraces current debates on the production of knowledge and its implications for the field in question. As a consequence, many natural science students (and scholars) in these fields have a rather rudimentary understanding of the research process both at a general level and with respect to their own doings. For example, many natural science students and scholars embrace the view that credible research is conducted according to the so-called "Scientific Method", often interpreted as Popper's idealized description of the hypothetic-deductive approach.[7] This view is highly problematic for several reasons. First, it suggests a widespread notion that most studies (not least your own) are conducted in line with the hypothetic-deductive approach, even though overwhelming evidence that this type of approach is rarely adhered to, especially not in empirically-based studies.[8] Second, it suggests that studies explicitly carried out by the use of other approaches than the hypothetic-deductive are by definition non-credible, which would imply that the majority of our empirical research traditions (also those in the natural sciences) are non-credible.

The fact that scholars from the natural sciences, engineering and medicine have at best a rudimentary understanding of the research process creates an imbalance when such issues are discussed in a heterogeneous group, putting the natural science scholars in a clearly disadvantageous position, as they, in the eyes of others, are deeply ignorant. This educational imbalance forms the basis of a hierarchy that is the reverse of the one discussed above in which qualitative (exact) natural sciences are ranked the highest, followed by quantitative (context bound) natural sciences, followed by quantitative social sciences and last by qualitative humanities.

Counteracting condescendance

A fruitful environment involves parties regarding each other as equals with skills, attributes and knowledge that will enhance the progress of and outcomes from the research. Two groups looking down their noses at each other is clearly not a fruitful environment. Breaking these hierarchies is necessary if we are to produce high-quality research across these traditional borders, and it is easier to break a hierarchy when you are on the top, rather than at the bottom.

First of all, if you belong to "the ignorant" side, you need to acknowledge that you lack competence in a crucial field and you need to get up to speed on a basic level. If you belong to the "knowledgeable" side you need to rid yourself of a potentially condescending attitude towards "the ignorant". Initiate dialogue-based discussion forums with fellow students and focus on your knowledge-producing practices. Use one or a couple of recent papers from your fields as the basis for a dialogue on knowledge-producing practices. Or, in more plain language: how is knowledge produced in this particular case? Make sure you understand pros and cons with quantitative as well as qualitative approaches.

Constructive collaboration demands avoiding the situation where those who are well-informed on overarching methodological issues look down upon those who have very little or no such training. In the flood of general literature on methodological issues, there are probably books that are better suited for some environments. If you have no such training, I encourage you to identify a few books or papers that are suitable for your context. Initiate a dialogue group with fellow students starting with your specific studies and discuss it in light of the questions in the framework presented in this book.

Humbleness and courage

The creation of a fruitful interdisciplinary environment lies in the ability to encourage humbleness, courage and self-confidence while suppressing arrogance. The arrogant researcher is the one who without hesitation uses procedures outside their own field of training. Interdisciplinarity is interpreted as "I am an experienced researcher, so I am capable of conducting any kind of research." The researcher who lacks in confidence on the other hand claims that one needs a certain type of undergraduate education to be capable of conducting certain types of studies. According to this view, you should only conduct studies that strictly lie within your own basic competence. Interdisciplinarity is interpreted as "I do what I was trained to do and you do what you were trained to do." Failure to acknowledge the experience and know-how required to gain competence on the specific level when crossing borders may result in bad research. Never stepping outside the practical realms of the skills once learned at the undergraduate level hinders border crossing and development. Neither position will render high-quality interdisciplinary studies.

Through a conscious approach you can build courage rather than arrogance and humbleness rather than self-abasement. To do so, you need to strengthen your meta-competence (simply put: your ability to think about your thinking). Academic training and experience can provide you with scholarly competence on a general level, not least of all in regard to research procedures. Chapters 6 to 11 deal with different facets of academic rigour. The crux is to learn to recognize that these facets are common for all scholarship – to distinguish the form from the inner kernel, to assess the relevance of a technique or method in light of the study as a whole, to view the specific in light of the general. The time needed to acquire the necessary skills to assess or conduct a certain study is considerably less for the experienced, open-minded researcher than for the undergraduate student. Many scholars who have been active for a while acquire a meta-competence, that is, an ability to critically assess a study on a general overarching level. By practising how to apply this critical approach to your own activities, it is possible to figure out what kind of specific competence you need in order to use a new approach or technique in a credible manner. The success of projects that bridge disciplines demands an awareness that all research procedures are based upon a general scholarly competence, as well as specific know-how and craftsmanship. By acquiring scholarly training that encourages and strengthens your ability to reflect upon the research and your learning process at an overarching level, you will enhance your ability to collaborate as well as your ability to produce high-quality work.

Outstanding studies

To produce stellar work is to do something original and to do it extremely well. It is undeniably more challenging to carry out studies that go beyond traditional academic disciplines, but those challenges are the very reason stellar work so often is interdisciplinary. To work outside your comfort zone, not shying away from areas where you need to be asking "why?" and jointly with other curious, unassuming minds, fearlessly pose questions that might be seen as plain stupid is a fast-track to original questions. The key is to combine such an unassuming, three-year-old attitude with a solid, rigorous approach to ensure that the work is not just original but also of superior quality. Dialogue and feedback is the key to both. When navigating the difficult terrain of interdisciplinary work, my hope is that the three dimensions of the framework presented in this book will help you produce stellar studies:

- **integration** of elements from different disciplines;
- **interaction** with organizations and individuals outside academia;
- **rigour** from an academic point of view.

Dialogue, feedback and how to manage supervisors

Finally – a few words on how to use dialogue-based groups and the relationship between supervisors and students. I have touched upon these issues several times throughout the book, so this is to sum up. As mentioned several times before, the governing idea of this book is to facilitate creation of interdisciplinary work by stimulating dialogues on quality and to draw on common-ground creation processes. Such dialogues bear double benefits: they enable identification and exploration of unknown and unexplored territories and they enable creation of high-quality work. Such dialogues will help you to identify, accept, respect and draw upon disciplinary-based cultural differences; in other words to find ways to *use* the differences. Good research comes in many forms and it is easier said than done to broaden your understanding of quality, but certainly more likely to happen if you strive to uncover, discuss and justify the explicit as well as the implicit norms guiding various perceptions of good vs. bad studies.

You can use dialogue-based discussions to increase your understanding on a general level, for example regarding different knowledge-producing practices and perceptions of quality. They can also be used to increase your awareness of your own perspective, what type of conclusions you may draw from what type of material and how design and context shape and influence the results and usefulness of a study. Such groups can also be used in a more direct manner to help you to clarify choices, make demarcations and explain the relevance of what you do. Another example is that they can be used to increase your awareness of your position in established

hierarchies and how that influences reflexes to defend, attack, be derogatory or just plain arrogant. It is crucial that you do not try to cover too
much ground in a meeting and that you have agreed why you meet and
what you will discuss. Using the questions of the framework in this book is
one way to get a focussed discussions going.

One of the challenges you face as a student is that some of the scholars
you meet will be unaware of the "beauty comes in many forms" principle.
It is possible that your supervisor or committee members belong to this
group. This will put you in a tricky situation. Your supervisor and committee
members are there to help and guide you. They are more experienced than
you and their advice is well worth listening to. But they might have a
narrow understanding of quality and how things "are supposed to be done".
It is crucial that you, your supervisor and your committee members have a
shared understanding of what you do and what the nature of your interdisciplinary studies is. It's worthwhile to bring the framework presented in
Chapter 3 to your supervisor's attention, especially Questions 1–5, and initiate a discussion on whether or not this could be a useful tool. If they don't
like the framework – ask them for something else. If they find all such
discussions irrelevant and time-consuming, find a group of students to
discuss the matters with and how to handle the problems. Your supervisor
and committee members are one of several assets at your disposal during
your studies – your job is to figure out how to make the best use of them.
With a creative approach, you will figure out how to frame the questions
you wish to discuss in a way that your supervisor and committee members
will find relevant and thus fruitful for you.

I wish you an exciting journey!

Notes

1   Enticed by the potential connection to *The Hitchhiker's Guide to the Galaxy* and the
    Don't Panic button
2   Primer is a word with multiple meanings, but they all carry the connotation of
    initiating action: an elementary book for teaching children to read or any book of
    elementary principles; something used as a first coat to make paint stick; or, as in
    molecular biology – a strand of nucleic acid that serves as a starting point for DNA
    replication.
3   See, for example, Tony Becher 1987a; b; Robert Birnbaum 1983.
4   See Michael Jay Katz 2006 *From Research to Manuscript*, Robert A. Day and Barbara
    Gastel 2006 *How to Write and Publish a Scientific Paper* or Björn Gustavii 2003 *How
    to Write and Illustrate a Scientific Paper*.
5   Such as, for example, *Basic Gas Chromatography* by Harold M. McNair and James
    M. Miller 1998; *GIS Tutorial* by Wilpen L. Gorr and Kristen S. Kurland 2007.

6   See, for example, the introduction in *Environmental Science, a Global Concern* by William Cunningham and Mary Cunningham 2005.
7   See, for example, Karl Popper *The Logic of Scientific Discovery* 2002 (Original *Logic der Forschung* 1935).
8   See, for example, Thomas Kuhn's *Structure of Scientific Revolutions* 1970; Bruno Latour's *Science in Action* 1987; Karin Knorr Cetina's *Epistemic Cultures: How the sciences make knowledge* 1999.

References

Primary sources

Bala, G., Calderia, K., Mirin, A., Wickett, M., Delire, C. and Phillips, T. J. (2006). Biogeophysical effects of CO_2 fertilization on global climate. *Tellus*, **58B**, 620–627.

Ban, N. C. (2008). Multiple perspectives for envisioning marine protected areas. A thesis submitted in partial fulfillment of the requirements for the degree of doctor of philosophy in the Faculty of Graduate Studies (Resource Management and Environmental Studies), The University of British Columbia (Vancouver). URI: https://circle.ubc.ca/handle/2429/1275 or http://en.scientificcommons.org/46175251 (M\y 9, 2010).

Bazermann, C. (1988). *Shaping Written Knowledge: The genre and activity of the written article in science*, The University of Wisconsin Press, Madison, WI.

Berg, K. (2003). Simulations of groundwater levels and soil water content: Development of a conceptual hydrological model with a continuous soil profile. Master thesis. Institutionen för tematisk utbildning och forskning, Miljövetarprogrammet, Linköpings universitet. ISRN LIU-ITUF/MV-D–03/13–SE.

Bohm, D. (1996). *On Dialogue*, Routledge, New York.

Campbell, D. T. (2005). Ethnocentrism of Disciplines and the Fish-Scale Model of Omniscience, in *Interdisciplinary Collaboration: An emerging cognitive science* (eds S. J. C. Derry, D. Schunn and M. A. Gernsbacher), Lawrence Erlbaum Associates, Inc., New Jersey.

Day, R. A. and Gastel, B. (2006). *How to Write and Publish a Scientific Paper*, Greenwood Press, Westport, CT.

Environmental Science & Policy Elsevier. Author instructions. http://www.elsevier.com/wps/find/journaldescription.cws_home/601264/authorinstructions (September 2007).

Environmental Science & Technology, American Chemical Society. Author instructions http://pubs.acs.org/paragonplus/submission/esthag/esthag_authguide.pdf (September 2007).

Eriksson, J., Frankki, S., Shchukarev, A. and Skyllberg, U. (2004). Binding of 2,4,6-trinitrotoluene, aniline, and nitrobenzene to dissolved and particulate soil organic matter. *Environmental Science & Technology*, **38** (11), 3074–3080.

Eriksson, J. (2003). Collective action and common effluent treatment plants: Field study within the textile industry Tirupur, South India. D-uppsats. Institutionen för tematisk utbildning och forskning, Miljövetarprogrammet, Linköpings universitet. ISRN LIU-ITUF/MV-D–03/16–SE.

Hagerman, S. (2009). Adapting conservation policy to the impacts of climate change: an integrated examination of ecological and social dimensions of change. A thesis submitted in partial fulfilment of the requirements for the degree of doctor of philosophy in the Faculty of Graduate Studies (Resource Management and Environmental Studies), The University of British Columbia (Vancouver). URI: https://

Interdisciplinary Environmental Studies: A Primer, 1st edition. © Gunilla Öberg.
Published 2011 by Blackwell Publishing Ltd.

circle.ubc.ca/dspace/handle/2429/7903 or http://en.scientificcommons.org/50560592 (May 9, 2010)

Holmstrand, H., Andersson, P. and Gustafsson, Ö. (2004). Chlorine isotope analysis of submicromole organochlorine samples by sealed tube combustion and thermal ionization mass spectrometry. *Analytical Chemistry*, **76**, 2336–2342.

Imo, R. (2007). Reflection paper. Graduate course RMES 502, Institute for Resources, Environment and Sustainability, University of British Columbia. With permission from the author.

Jägerskog, A. (2003). *Why states cooperate over shared water: The water negotiations in the Jordan River Basin*. Doctoral thesis. Linköping Studies in Arts and Science #281. Institutionen för Tema, avd. för Vatten i natur och samhälle, Linköpings universitet. ISBN 9173737496 ISSN 02829800.

Jones, K. (2007). Reflection paper. Graduate course RMES 502, Institute for Resources, Environment and Sustainability, University of British Columbia. With permission from the author.

Karlsson, S. (2003). Baltic Sea environmental co-operation – a Swedish perspective on agricultural discharge issues within HELCOM and Baltic 21. Master thesis. Institutionen för tematisk utbildning och forskning, Miljövetarprogrammet, Linköpings universitet. ISRN LIU-ITUF/MV-D-03/22-SE.

Kauffman, S. J., Royer, D. L., Chang, S. and Berner, R. A. (2003). Export of chloride after clear-cutting in the Hubbard Brook sandbox experiment. *Biogeochemistry*, **63**, 23–33.

Kirchner, J. W., Feng, X. and Neal, C. (2000). Fractal stream chemistry and its implications for contaminant transport in catchments. *Nature*, **403**, 524–527.

Lahsen, M. (2004). Transnational locals: Brazilian scientists in the climate regime. I: Earthly politics, in *Local and Global in Environmental Governance* (eds S. Jasanhoff. and Long Martello, M. (red.)), MIT Press, Cambridge, MA. ISBN 026210103, pp. 151–173.

Larsson, E. (2004). Science and policy in the international framing of the climate change issue. Master thesis. Institutionen för tematisk utbildning och forskning, Miljövetarprogrammet, Linköpings universitet. ISRN LIU-ITUF/MV-D-04/11-SE.

Lepenies, W. (2001). Rektor at the Institute for Advanced Study in Berlin, quoted on the back cover of *Re-Thinking Science: Knowledge and the public in an age of uncertainty* (eds H. Nowotny, P. Scott, *et al.*), Polity Press, Cambridge.

Lindgren, T. (2002). *Hash: Overlook Duckworth*, Peter Mayer Publishers, New York.

Lister, J. (2009). Co-regulating corporate social responsibility: government response to forest certification in Canada, the United States and Sweden. A thesis submitted in partial fulfilment of the requirements for the degree of doctor of philosophy in the Faculty of Graduate Studies (Resource Management and Environmental Studies), The University of British Columbia (Vancouver). URI: https://circle.ubc.ca/handle/2429/7212 or http://en.scientificcommons.org/50560753 (May 9, 2010)

Lister, J. (2007). The myth, reality and social process of sustainable forest management, in *Sustainable Resource Management: Reality or Illusion?* (ed. P. N. Nemetz), Edward Elgar, London.

Lister, J. (May 2008). E-mail communication. Reprinted with permission from author.

Liu, Y., Guo, H., Zhang, Z., Wang, L., Dai, Y. and Fan, Y. (2007). An optimization method based on scenario analysis for watershed management under uncertainty. *Environmental Management*, **39**, 678–690.

Lundén, A. (2003). Biogas production: anaerobic digestion of grains diluted in process water from a wastewater treatment plant. Master of Science thesis, Environmental Science Programme. Linköping University, Department of Thematic Studies. ISRN: LIU-ITUF/MV-D-03/15-SE.

Müller, C., Stevens, R. J., Laughlin, R. J., Ottow, J. C. G. and Jäger, H. -J. (2003). Ammonium immobilisation during chloroform fumigation. *Soil Biology and Biochemistry*, **35**, 651–665.

Nilsson, A. E. (2007). A changing arctic climate: science and policy in the arctic climate impact assessment. Doctoral thesis. Linköping University, The Tema Institute, Department of Water and Environmental Studies. ISBN: 978-91-85715-23-7.

Nowotny, H., Scott, P. and Gibbons, M. (2001). *Re-thinking science: Knowledge and the public in an age of uncertainty*, Polity Press, Cambridge.

Öberg, G., Holm, M., Parikka, M., Sandén, P. and Svensson, T. (2005). The role of organic-matter-bound chlorine in the chlorine cycle: a case study of the Stubbetorp catchment, Sweden. *Biogeochemistry*, **75**, 173–201.

Oreskes, N. (2004). Science and public policy: what's proof got to do with it? *Environmental Science & Policy*, **7**, 369–383.

Ortiz-Bermúdez, P., Srebotnik, E. and Hammel, K. E. (2003). Chlorination and cleavage of lignin structures by fungal chloroperoxidases. *Applied and Environmental Microbiology*, **69**, 5015–5018.

Osgyani, P. (2003). The seasonal variation in nitrate leakage from a clear-cut spruce forest in the province of Halland, Sweden. Master thesis. Institutionen för tematisk utbildning och forskning, Miljövetarprogrammet, Linköpings universitet. ISRN LIU ITUF/MVD-03/03—SE.

Pielke, R. A. J. (2007). *The honest broker: Making sense of science in policy and politics*, Cambridge University Press, Cambridge.

Reina, R. G., Leri, A. C. and Myneni, S. C. B. (2004). Cl k-edge X-ray spectroscopic investigation of enzymatic formation of organochlorines in weathering plant material. *Environmental Science & Technology*, **38**, 783–789.

Soil Biology & Biochemistry, Elsevier. Author Instructions. http://www.elsevier.com/wps/find/journaldescription.cws_home/332/authorinstruction (September 2007).

Weiss, S. (2003). Illegal logging as a potential source of conflicts: A case study of the Indonesian forestry sector. Master thesis. Institutionen för tematisk utbildning och forskning, Miljövetarprogrammet, Linköpings universitet. ISRN LIU-ITUF/MV-D-03/25—SE.

White, L. Jr. (1974). The ecological roots of our ecological crisis, in *Ecology and Religion in History* (eds D. Spring and E. Spring), Harper Torchbooks. Harper & Row, Publishers. New York, Hagerstown, San Francisco, London, pp. 15–31.

Wilson, C. (2008). Understanding and influencing energy efficient renovation decisions. A thesis submitted in partial fulfilment of the requirements for the degree of doctor of philosophy in the Faculty of Graduate Studies (Resource Management and Environmental Studies), The University of British Columbia (Vancouver). URI: https://circle.ubc.ca/handle/2429/2388 or http://en.scientificcommons.org/46174194 (May 9, 2010).

Wesström, K. (2002). Spatial patterns and storage of organic chlorine and chloride in coniferous forest soil in southeast Sweden. BSc Honors thesis. Institutionen för tematisk utbildning och forskning, Miljövetarprogrammet, Linköpings universitet. ISRN LIU-ITUF/MV-C-02/16—SE.

Secondary sources

Apostel, L. (1972). Conceptual tools for interdisciplinarity: An operational approach, in *Interdisciplinarity: Problems of teaching and research in universities*, OECD, Centre for Educational Research and Innovation, Paris.

Arnstein, S. R. (1969) Ladder of citizen participation. *Journal of the American Institute of Planner*, **4**, 216–224.

Barry, A., Born, G. and Weszkalnys, G. (2008). Logics of interdisciplinarity. *Economy and Society*, **37**, 20–49.

Barth, R., and Steck, R. (eds) (1979). *Interdisciplinary Research Groups: Their management and organization*, Interstudy, Seattle.

Bazerman, C. (1988). *Shaping Written Knowledge: The genre and activity of the experimental article in science*, The University of Wisconsin Press, Madison, WI.

Becher, T. (1987a). The disciplinary shaping of the profession, in *The Academic Profession: National, disciplinary and institutional settings* (ed. B. R. Clark), University of California Press, Berkeley, CA., pp. 271–303.

Becher, T. (1987b). Disciplinary discourse. *Studies in Higher Education*, **12**, 261–274.

Birnbaum, R. (1983). *Maintaining Diversity in Higher Education*, Jossey-Bass, San Francisco, CA.

Bohm, D. (1996). *On Dialogue*, Routledge, New York.

Boix, M. V. (2004). Assessing student work at disciplinary crossroads. in *GoodWork® Project Report Series*, Harvard University Press, Cambridge, MA.

Boix, M. V. and Gardner, H. (2003). Assessing interdisciplinary work at the frontier: An empirical exploration of "symptoms of quality", in *GoodWork® Project Report Series*, Harvard University Press, Cambridge, MA.

Bradbaer, J. (1999). Barriers to interdisciplinarity – disciplinary discourses and student learning. *Journal in Geography in Higher Education*, **23**, 381–396.

Campbell, D. T. (2005). Ethnocentrism of disciplines and the fish-scale model of omniscience, in *Interdiscipinary Collaboration: An emerging cognitive science* (eds S. J. Derry, C. D. Schunn. and M. A. Gernsbacher), Lawrence Erlbaum Associates, Inc., New Jersey.

Cotterell, R. B. M. (1979). Interdisciplinarity: The expansion of knowledge. *Higher Education Review*, **11**, 47–56.

Cunningham, W. and Cunningham, M. (2005). *Environmental Science, A Global Concern*, Canadian edn, McGraw Hill-Ryerson, Toronto.

Daston, L. and Galison, P. (2007). *Objectivity*, New York: Zone Books.

Day, R. A., and Gastel, B. (2006). *How to Write and Publish a Scientific Paper*, Greenwood Press, Westport, CT.

Denzin, N. K. and Lincoln, Y. S. (2005). *The SAGE handbook of Qualitative Research*: Sage Publications Inc., Thousand Oaks, CA.

Derry, S. J., Schunn, C. D. and Gernsbacher, M. A. (2005). *Interdisciplinary Collaboration*, Lawrence Erlbaum Associates, Inc., New Jersey.

Field, M., Lee, R. and Field, M. L. (1994). Assessing interdisciplinary learning. *New Directions for Teaching and Learning*, **58**, 69–84.

Fleck, L. (1979) *Genesis and Development of a Scientific Fact*, University of Chicago Press, Chicago.

Frodeman, R. and Mitcham, C. (2003). The new directions initiative: Integrating the humanities into science-society relations. *Quarterly Colorado School of Mines*, **103**, 5–16.

Frodeman, R., Klein, J. T. and Mitcham, C. (eds) (2010). *The Oxford Handbook of Interdisciplinarity*, Oxford University Press, Oxford.

Funtowics, S. O. and Ravetz, J. R. (1993). Science for the post normal age. *Futures*, **25**, 735–755.

Gibbons, M., Limoges, C., Nowotny, H., Schwartzman, S., Scott, P. and Trow, M. (1994). *The New Production of Knowledge: The dynamics of science and research in contemporary societies*, Sage Publications, Inc., Thousand Oaks, CA.

Giltrow, J. (2002). *Academic Writing: Writing and Reading across the Disciplines*, 3rd edn, Broadview Press, Peterborough, Ont.

Gorr, W. R. and Kurland, K. S. (2005). *GIS Tutorial: Workbook for ArcView 9.0*: ESRI Press; 2nd edn.

Gustavii, B. (2003). *How to Write and Illustrate a Scientific Paper*, Cambridge University Press, Cambridge.

Haraway, D. (2008). *When Species Meet*, University of Minnesota Press, Minneapolis, MN.

Heckhausen, H. (1972). Discipline and interdisciplinarity, in *Interdisciplinarity: Problems for teaching and research in universities*, OECD, Centre for Educational Research and Innovation, Paris, pp. 83–89.

Heintz, C., Origgi, G. and Sperber, D. (2007). Rethinking interdisciplinarity. URL http://www.interdisciplines.org/interdisciplinarity (May 9, 2010).

Hughes, T. P. (1998). *Rescuing Prometheus*, Pantheon Books, New York.

Jantz, E. (1972). Inter- and transdisciplinary university: A systems approach to education and innovation. *Higher Education*, 1, 7–27.

Jasanoff, S. (1998). *The Fifth Branch. Science advisors as policy makers*, Harvard University Press, Cambridge, MA.

Kast, F. E. and Rosenzweig, J. E. (1970). Interdisciplinary programs in a university setting. *Academy of Management Journal*, 13, 311–324.

Katz, M. J. (2006). *From Research to manuscript: A guide to scientific writing*, Springer, Dordrecht, NL.

Knorr Cetina, K. D. (1999). *Epistemic Cultures: How the sciences make knowledge*, Harvard University Press, Cambridge, MA.

Kuhn, T. S. (1970). *The Structure of Scientific Revolutions*, University of Chicago Press, Chicago, IL.

Kurland, K. S. and Gorr, W. R. (2007). *GIS Tutorial for Health*, ESRI Press, New York.

Kvale, S. (1996) *InterViews: An Introduction to Qualitative Research Interviewing*, Sage Publications Inc., Thousand Oaks, CA.

Latour, B. (1988). *Science in Action*, Harvard University Press, Cambridge, MA.

Lattuca, L. R. (2001). *Creating Interdisciplinarity: Interdisciplinary Research and Teaching among College and University Faculty*, Vanderbilt University Press, Nashville, TN.

Luszki, M. B. (1958). *Interdisciplinary Team Research: Methods and Problems*, New York University Press, New York.

Mallery, J. C., Hurwitz, R. and Duffy, G. (1986). Hermeneutics: From textual explication to computer understanding? in *Artificial Intellegence Memo*, Artificial Intelligence Laboratory, MIT, Cambridge, MA.

McNair, H. M. and Miller, J. M. (1998). *Basic Gas Chromatography*, John Wiley & Sons, New York.

Mothe, J. D. L. (1992). *C.P. Snow and the Struggle of Modernity*, McGill-Queen's University Press, Montreal.

Mushakoji, K. (1978). Peace Research as an International Learning Process: A New Meta-Paradigm. *International Studies Quarterly*, 22, 173–194.

Nowotny, H., Scott, P. and Gibbons, M. (2001). *Re-thinking science. Knowledge and the public in an age of uncertainty*, Polity Press, Cambridge.

O'Donnell, A. M. and Donnell, S. J. (2005). Cognitive processes in interdisciplinary groups: Problems and possibilities, in *Interdisciplinary Collaboration* (eds S. J. Derry, C. D. Schunn and M. A. Aernbacher), Lawrence Erlbaum Associates, Inc., Mahwah, NJ.

Öberg, G. (2002). The natural chlorine cycle – fitting the scattered pieces. *Applied Microbiology and Biotechnology*, 58, 565–581.

Öberg, G. (2007). *Praktisk tvärvetenskap – tankar om och för gränsöverskridande projekt (in Swedish). Interdisciplinarity in practice: Thoughts concerning boundary crossing projects*, Studentlitteratur, Lund.

Öberg, G. (2009). Facilitating interdisciplinary work: using quality assessment to create common ground. *Higher Education*, **57**, 405–415.

Pielke, R. A. J. (2007). *The Honest Broker: Making sense of science in policy and politics*, Cambridge University Press, Cambridge.

Popper, K. (2002). *The Logic of Scientific Discovery*, Routledge, New York.

Repko, A. (2008). *Interdisciplinary Research. Process and Theory*, Sage Publications, Inc., Thousand Oaks, CA.

Robinson, J. (2008). Being undisciplined: Transgressions and intersections in academia and beyond. *Futures*, **40**, 1, 70–86.

Salter, L. and Hearn, A. (1996). *Outside the Lines: Issues in interdisciplinary research*, McGill-Queen's University Press, Montreal.

Sarewitz, D., Pielke, R. A. and Byerly, R. (eds) (2002). *Prediction: Science, decision making, and the future of nature*, Island Press, Washington, DC.

Shatz, D. (2004). *Peer Review: A critical inquiry*: Rowman & Littlefield Publishers, Inc., Lanham, Md.

Snow, C. P. (1993). *The Two Cultures*, Cambridge University Press, Cambridge.

Strand, R. (2002). Complexity, Ideology and Governance. *Emergence*, **4**, 164–183.

Sung, N. S., Gordon, J. I., Rose, G. D., Getzoff, E. D., Kron, S. J., Mumford, D. *et al.* (2003). Educating future scientists. *Science*, **301**, 1485.

Thompson Klein, J. (1990). *Interdisciplinarity: History, theory and practice*. Wayne State University Press, Detroit, MI.

Thompson Klein, J. (1996). *Crossing Boundaries: Knowledge, disciplinarities and inter-disciplinarities*, University Press of Virginia, Charlottesville, VA.

Walter, A., Helgenberger, S., Wiek, A. and R. W. S. (2007). Measuring social effects of transdisciplinary research – Design and application of an evaluation method. *Evaluation and Program Planning*, **30**, 325–338.

Wiek, A. (2007). Challenges of Transdisciplinary Research as Interactive Knowledge Generation – Experiences from Transdisciplinary Case Study Research. *Gaia – Ecological Perspectives for Science and Society*, **16**, 52–57.

Index

Interdisciplinary Environmental Studies: A Primer, 1st edition. © Gunilla Öberg.
Published 2011 by Blackwell Publishing Ltd.

Index compiled by Terry Halliday